Weather Flying

Also by Robert N. Buck

FLYING KNOW HOW
THE ART OF FLYING
THE PILOT'S BURDEN

Other McGraw-Hill Aviation Titles

Wolfgang Langewiesche
STICK AND RUDDER: AN EXPLANATION OF THE ART OF FLYING

Richard L. Taylor
INSTRUMENT FLYING, FOURTH EDITION

Tony T. Kern
REDEFINING AIRMANSHIP

Lewis Bjork
PILOTING FOR MAXIMUM PERFORMANCE

John F. Welch, Editor
VANSICKLE'S MODERN AIRMANSHIP

Paul Dempsey, Andrew Goetz, and Joseph Szyliowicz
DENVER INTERNATIONAL AIRPORT: LESSONS LEARNED

Weather Flying

Robert N. Buck

Fourth Edition

McGraw-Hill

New York   San Francisco   Washington, D.C.   Auckland   Bogotá
Caracas   Lisbon   London   Madrid   Mexico City   Milan
Montreal   New Delhi   San Juan   Singapore
Sydney   Tokyo   Toronto

McGraw-Hill

A Division of The *McGraw·Hill* Companies

15 16 17 18 19 20   QWF/QWF   09

ISBN 0-07-008761-X

The sponsoring editor for this book was Shelley Ingram Carr, the editing supervisor was Sally Glover, and the production supervisor was Claire Stanley. It was set in Garamond by McGraw-Hill's desktop publishing department in Hightstown, N.J.

Printed and bound by Quebecor/Fairfield.

McGraw-Hill books are available at special quantity discounts to use as premiums and sales promotions, or for use in corporate training programs. For more information, please write to the Director of Special Sales, McGraw-Hill, 11 West 19th Street, New York, NY 10011. Or contact your local bookstore.

GEN1-AV2

For Leighton Collins.

About the Author

Leading aviation author Robert N. Buck (Fayston, Vt.) set a New York to Los Angeles speed record at the age of 16. A retired TWA jet captain, Buck has flown the Atlantic more than 2,000 times. As a civillian, he headed a four-year bad weather research project for the Air Force, which won him an Air Medal. Buck has been consultant to four FAA Administrations and airlines on many aspects of aviation safety, and is the author of *The Art of Flying* and *Flying Know-How*, among others. An avid general aviation pilot throughout his life, Buck currently owns a Schleicher ASW-20 high performance sailplane with his son.

Contents

About Some People

There are many people to thank for helping me with this book. It starts with a meteorologist named Mr. Ball, who, when I was sixteen and starting across the United States alone, briefed me at Newark Airport, New Jersey, on what the weather would be. I was interested in what he said because I didn't know much about weather, or flying either, and I was a little nervous.

Mr. Ball—I never knew his first name—also awakened my interest in weather, what makes it, and what a pilot does about it. Looking back, I realized that Mr. Ball was an exceptionally good meteorologist.

After Mr. Ball came a host of meteorologists, most of whose names I don't know and never will. They are voices and faces over a telephone or in a weather office in some city when I was trying to make cross-country records, or just going somewhere.

Some time later they were in wood-paneled rooms in Gander, Newfoundland, or Prestwick, Scotland, during World War II when the Atlantic, sitting out there waiting to be crossed, was a big, black unknown, and whatever that man said I listened to carefully because any crumb of information was something to grab for help in doing a job that I didn't know all there was to know about doing.

There are meteorologists I got to know well by searching conversation as I tried to learn, and later in consultation as we probed the atmosphere in research—men like Ed Minser and J.A. Browne of TWA. I continued to learn from TWA meteorologists who briefed me on what the weather would be over the North Atlantic, which became an ocean I respected but felt very comfortable flying over in a 747.

There is Nieut Lieurance (retired) of the National Oceanic and Atmospheric Administration (NOAA). Nieut is a pilot's meteorologist and has done much to bring good weather services to us all. Chuck Lindsay (retired) of NOAA is one of the world's foremost soaring meteorologists, and a person always good for a new piece of knowledge. I thank Chuck also for reading the manuscript of the first edition and making some valuable suggestions.

Lt. General Ira C. Eaker, a boyhood hero of mine, saw the possibilities of thunderstorm research and made it possible for TWA and the Air Force to work together using a P-61 Black Widow night fighter which I was fortunate enough to fly in the project.

Captain Robert A. Wittke, TWA (retired), my close friend since childhood, made useful and perceptive suggestions.

Also, Dan Sowa, the superb meteorologist of Northwest Airlines, has given me many insights into the mysteries of weather, as has his associate, the brilliant technical airman, Captain Paul A. Soderlind of Northwest Airlines.

I'm indebted to my good friends at the National Weather Service in Burlington, Vermont, and the staff at the Flight Service Station in Burlington, Vermont. These men and women have been ready and most willing to patiently answer my questions and enthusiastically dig out information for me. I especially thank Ben Martin, Betty Czina, and Jimmy Mason of the FSS and William Grady of the Burlington NWS.

Years ago I learned from Dr. Horace Byers and for a brief but exciting and privileged time from Dr. Irving Langmuir.

I wish I could name all the others, but there isn't space. I do want it to be clear, however, that I thank all meteorologists. They are a misunderstood group who are cussed more than praised, when it should be the other way around.

I want to thank Velma and Dwane Wallace for urging me to get this book done.

Solid advice and encouragement came from Charles W. Ferguson and Caroline Rogers of *The Reader's Digest*. They were patient and understanding friends.

I must mention the man to whom this book is dedicated, Leighton Collins, who with his *Air Facts* magazine started me writing about weather almost 50 years ago. His enthusiasm and encouragement did the most.

My wife Jean, daughter Ferris, and son Rob, captain on a major airline and an experienced weather pilot who shares my weather interest and study, all gave their enthusiastic support, and I thank them for that.

Robert N. Buck
Fayston, Vermont

Introduction to the First Edition

What Captain Robert Buck says in *Weather Flying* has not been said before. Other books explain how weather is made; this book explains how weather is flown.

We get to the field in the morning. Here is the weather: the map, the forecasts, the sequence of current reports from many points, the winds aloft—the whole package, prepared by experts. However deep our knowledge of meteorology may be, we cannot hope to do better. For us as pilots, the question is: What do I *do* with this? Go or no go? If go: go underneath the weather, or on top, or through? Or go around it? Follow a railroad? File an instrument flight plan? Go right away? Delay a couple of hours?

Those questions. They are the last of the real problems of flying. Everything else has now been quite well mastered. The airplane itself now works: it handles nicely (at least, those flown by the general public). It has climb to spare, and we can usually find some level where the air is smooth. Noise and vibration are subdued: we can stand long hours in a day and make big distances. Electronics tells us where we are. Airports are plentiful and runways long. Engine failure is so rare we now almost forget about it. Even the economics of flying are no longer so forbidding. But the weather...

It is not often clearly impossible. When it is, we have no problem: back to the hotel. On the contrary (as Captain Buck points out), the weather is normally flyable. Again, no problem, or only easy ones. But every once in a while—depending on season and part of the world—something is sitting out there that worries us. If you fly far enough in a straight line, you're likely to come up against some problem weather that very day. How well we deal with those situations determines how well the airplane works.

Too bold, and we cause emergencies and have accidents. Too timid, and we destroy the utility of the airplane and let our skill as pilots atrophy. Then pretty soon we have to be more timid still.

These weather decisions can be painful to make because we don't really know how to go about making them. And we know we don't know! We make them often by a process which is a dumb, confused struggle between "guts" and "judgment," ambition and fear. How it comes out depends on how we felt that day when we last had a good scare, whether the girlfriend is looking, or "This town is full, there is a convention on, you can't get a hotel room, let's go." Things like that tip the balance. And so (in Buck's words) we "drag the luggage back to town," often for no reasoned cause, or we "fling ourselves into the air," often with only a vague estimate of what's ahead. Or else, more likely, we stand around for another hour, look at the weather map some more, wait for the next weather sequence to come on the teletype and think, "I wish I could talk this over with some really experienced friend." That's where Bob Buck comes in.

Experienced? We could consider 2,000 hours quite a respectable lot of time. Buck has some 2,000 Atlantic crossings! And those are only half of his experience. At this writing, Captain Buck's log records well over 29,000 hours, all types. He set his first record when he was 16, in a Pitcairn Mailwing, an open-cockpit biplane. It was the Junior Transcontinental Record, New York to Los Angeles, and "It didn't amount to much—mostly just getting there." He set another record, for nonstop distance in light airplanes, in a 90-horsepower Monocoupe, overnight from California to Ohio—his engine quit there. With the same airplane he joined an expedition and searched the jungle of Yucatan for Maya ruins. He joined TWA as a co-pilot at the time when the DC2 was the Giant Airliner, then the DC3.

In those ships, between Pittsburgh and Newark, on a winter night, a pilot could learn a lot about flying the weather. It was about the toughest weather flying ever done. Instrument flying was new then. Radio aids and instruments were still quite crude, airplanes comparable in performance to present-day private airplanes. But the airlines flew almost the same weather as now. TWA had a milk run from Kansas City to Newark that took all night and made nine stops, and some nights all nine required an instrument approach.

TWA made Bob Buck a captain in two years. Then the war came and the first great surge of ocean flying, in four-engined land planes, under the aegis of Air Transport Command. Buck became Assistant Director of Training for TWA's Intercontinental Division (which was part and parcel of ATC). He checked people out on the DC4s and on the ocean routes. Next, for four years, he was the captain and manager of a special research project that TWA had taken on for the Army Air Forces. He had a B-17 bomber ("Flying Fortress") to himself and a mission to seek out precisely the kind of weather that others stayed away from—the kind that gave the most trouble. It started with research into snow static—the kind you find best in Alaska. Icing research was added, then other things; before long, he was carrying 14 different projects. To find enough weather that was bad enough, Buck ranged from Alaska to Panama, out to Hawaii, to Southeast Asia, and, once, clear around the world. He ended up doing outright thunderstorm research. For this, they gave him a P61 twin-engined fighter (the "Black Widow") with lots of radar and lots of structural strength. He flew thunderstorms forward and backward, slow and fast, high and low. What was inside the monsters? Could you get through without spilling your gyros? How much did you dare slow up the airplane? What was the best way to keep control of the airplane? How bad was lightning, hail, turbulence? It was one of the first deliberate, systematic series of flights through thunderstorms ever undertaken.

Buck then went back on the line, but kept doing new flight research for the military and the airlines both—on airborne radar and on *low*, low instrument approaches. When the Instrument Landing System (ILS) first came in, he was one of those who had to figure out the best way to use it. Once he wanted to study runway visibility in fog as it appears to a pilot breaking out of a low overcast. He let himself be hoisted up into the overcast, hanging in a parachute harness from a captive balloon. He likes to fly.

He likes that left-hand front seat of an airplane. At one time they made him chief pilot of TWA. He found that the desk work interfered with flying, and he quit and went back on the line as a captain. "The sky is my office," he once wrote. He was again tapped for the executive side of the company. TWA's president at the time talked to him about it and held out the prospect of a vice-presidency. Buck said: "Mr. Burgess, there are only two jobs on this airline I would want—yours or mine."

He owns a high-performance sailplane, which he flies in cross-country competition. He once made a private tour of Africa in a DC3. He has been around the world once sidewise—i.e., via both poles. He sits on various national and international committees that deal with piloting, is active in the Air Line Pilots Association, and acts as a consultant to manufacturers of airplanes and electronics. But always, when you see him again, he is just back from Paris or Bombay, with a cabinful of passengers riding behind him.

So that's the man who now comes in, as we stand there studying the weather map and debating what to do. He knows our problem. We are not engaged in an academic exercise: We are making, or passing up, a serious commitment. As we search the weather map we are also searching our souls: Am I good enough for this? Can I hack it? On paper, this weather situation can be dealt with by such-and-such a procedure. In reality, when the pressure comes on, will I get flustered and panicky, so that I can't do my best? Airplanes are flown in weather by real people, and the pilots—we, ourselves—are part of the situation.

Buck himself has been through enough tense situations to—well, to write a book. He's seen fellows' knuckles turn white on the controls—and maybe sometimes his own. There are some interesting short passages in his book where he touches on that side—panic control, self-control, the *constructive* use of the imagination. You can keep an easy touch on the control, he says, even though your knees are shaking. These passages will convey a little of the man who is talking to us—his air of ease, cultivated by long self-training and mental discipline. They will also give you the impression that he is your friend. It's interesting to note here some things he does *not* say.

He does not bore us, for example, with the phony good advice, "Don't Exceed Your Limitations..." We do not know exactly what our limitations are. You don't know the breaking strength of a material until you have broken it. Nor do we know how tough this particular weather situation will be to fly. That's just our problem!

And he does not come at us with an advanced course in meteorology. Certainly it helps, in flying the weather, to know some textbook meteorology, to have clear concepts of the things we meet in the air: cold fronts and warm fronts, cumulus clouds and thunderstorms, inversions, dewpoint and temperature, and types of fog, the principal air masses in our part of the world. This knowledge helps us understand the language of the weather person and the meaning of the weather map. And it helps us to recognize these weather phenomena when we meet them. The nature of the brain is such that we see what we have seen before, and what we have a name for. We are blind to things which have not been properly introduced. People had fronts passing over them for thousands of years, but nobody ever saw a front as *front*—i.e., as boundary between contrasting air masses. Then, 50 years ago, the Norwegians first recognized the *cold front*, described it, named it. Now everybody can plainly see many frontal passages every year. In this sense, some descriptive meteorology helps the pilot fly the weather. But with the pilot's main problem—

What do I *do?*—textbook meteorology helps only little. And just because it helps so little we are tempted to give him more—more than he has any use for: to go deep into the question of how weather is made. The coriolis force, the geostrophic wind, the latent heat of condensation, the adiabatic lapse rate, frontogenesis and frontolysis. All that is a fascinating look into God's kitchen, but the pilot does not want to make the weather—or even the forecast. All he wants to do is fly it. And for that he does not need *more* meteorology; he needs a different kind and he is getting it here.

What is it Captain Buck does for us? It's like untying a knot. The bafflement we feel as we try to judge a weather situation is a sort of knot in which everything is balled together. Weather. He shows us where to find the place to start and how to unravel it.

Read what he says about the Big Picture, the If-Thinking, and Way Out, and right away the problem takes on order. These are indeed things on which we can ask questions, find the information, and make judgment. Perspectives open, a strategy suggests itself. More questions follow. What will be the influence of the local terrain, the time of day? You feed that in, and you can make more judgments. Now you have more to judge by: This mountain range is okay now but probably will have many thunderstorms in the afternoon. Better go now. This city with this wind direction will have industrial smoke in the morning, but visibility will improve by noon; you gain by delay. And so on. By showing us what the productive questions are, Captain Buck arms us with a judgment capability we never knew we had.

Another problem that baffles us: how to monitor the weather once we are en route. The weather here and now is okay. But where I'm going it is different, and by the time I get there it will be different again. Is the situation solid? Should I do something or just keep flying? How can I know?

Sure, we can get weather information by radio. But the same old problem: What is the intelligent question to ask? And what do we *do* with the information? Should we do anything? Many of us find it difficult to get an effective thought process started. We just keep on flying and hope the weather will hold. Buck reminds us that a whole weather system may not move as fast as "they" expect, or may move faster or may even back up. This is perhaps the most frequent reason for forecasts going sour. He thereby gives us another productive question to ask: Is the weather system moving as expected?

Now that we see the question, we can often get the answer we need by watching (for example, the progress of a cold front) at points which may be quite far off to the side of our route. It's really a quite simple idea. It's just a sample the captain gives us of what goes on in his mind when he is captain. But it greatly smartens us up.

Bob Buck writes as he talks and flies, with an easy touch. He uses small words. Unlike most professionals, he does not try to make his art seem mysterious and difficult: He makes it seem simple and plain. This might fool some reader into thinking that he is getting just a light chat. Not so. What we get here is in reality a sophisticated course in problem solving. Buck shows us not *what* to think but *how* to think. Not "What should I do?" but "How do I go about deciding what to do?"

Rules never quite fit the real-life situation. But a man who acts on rational grounds and knows what his reasons are can deal with the variety of real-life situations realistically. And should things *not* work out as expected, should the judgments have been mistaken or the information false, such a man discovers his error early, while he still can do something about it. Rational decision making: In business or government, we call this the Harvard Business School Approach. It is an American specialty, and it has been immensely productive. It will be productive for readers of this book.

Because of this sophisticated approach, Bob Buck's book is equally useful to pilots of all experience levels. All pilots do not fly the same degree of weather. Many people have to fly Visual Flight Rules (VFR); some fly instruments. Many have to be cautious; some can fly tough. But all have much the same problem of decision making—the problem the captain shows us how to solve. And so, this book being about this sort of thing, and written by this sort of man, a pilot needs an introduction to it like a hungry man needs an introduction to a steak. Just start right in.

Wolfgang Langewiesche
Princeton, New Jersey

1

Weather Flying

Weather bothers our flying only in a few basic ways; it prevents us from seeing; it bounces us around to the extent that it may be difficult to keep the airplane under control and in one piece; and ice, wind, or large temperature variations may reduce the airplane's performance to a serious degree.

That's what weather does. There are degrees and nuances, but all in all, we fight weather in order to see, to keep under control, and to get the best and safest performance from an aircraft. The question is, "How?"

Well, we should know something about weather, what it is made of, and how it moves. But a pilot wants to know practical things, what to do about the weather and how to cope with its capriciousness. These practical things are a philosophy for thinking about weather and methods for flying it.

Prior to developing a philosophy and methods, a pilot should have one point firmly etched in mind: weather forecasting is not an exact science. This statement is an old one, but it's true and ought to be thought about before tossing it aside as old hat.

The best weather person or an impersonal computer cannot forecast with perfect accuracy. The National Weather Service can make impressive statements about how well they do, and their numbers are valid and true. In the overall picture, they do a good job, percentagewise; but the time they miss and pull their accuracy down from 100 percent to something even a little less may happen to be the night our destination was forecast clear, but fell on its face to zero-zero! Right then the pilot isn't much interested in statistics—except in trying not to become one.

This shouldn't be taken as criticism of meteorologists or computers. Actually, a compassionate, understanding, and friendly feeling toward them will do the most good, but understand the cold fact; you cannot count on weather always doing what is forecast because even with fancy computer models, satellites, and perhaps a little witchcraft, we simply cannot outguess it 100 percent of the time. This means that there will be times when the weather will not work out as forecast, when it will be bad rather than good. This is a fact of flying life and we must always be prepared for it, accepting it as part of the game, ready to cope with it coolly and free of emotion. And that is the most important statement in the book!

The pilot's weather philosophy has two parts. The first is skepticism. Being a weather skeptic is an important ingredient of the formula for living to a ripe old age. The second part is always to have an alternate plan of action. These two keys, skepticism and alternate action, are the foundation of it all.

Being a skeptic keeps us safe; having an alternate plan of action adds to safety, but more important it makes it possible to fly and to make the airplane work.

If we are completely skeptical, we put the airplane in the hangar and forget about it, and sometimes this is a good idea. But we are trying to use airplanes; we want to go places as much as possible. The alternate plan of action helps us to do so.

Say a pilot is headed for a place that's forecast clear; skepticism says it's near the coast, night is approaching, the temperature is near the dewpoint, and the wind from the sea—our destination airport could fog in. Now the easiest way out would be to stay home: but we could also take off and probably make it, as long as there is enough fuel to get away from the coast, should it fog in, and fly back inland to a fog-free airport. That's alternate action.

That is a very simple example, but it is basic and important. It is the way the airlines operate and the way they keep going. They are skeptical about weather—skeptical enough to have an alternate plan of action in everything they do.

To fly in weather a pilot needs a certain ability and various degrees of equipment. Since all pilots cannot have all the ability and all the equipment, they must take on weather in amounts that fit ability and equipment. This can be done, and the fact that a pilot doesn't have instruments, radio, and a rating to use them doesn't mean he or she has to stay on the ground whenever there is a cloud in the sky. But pilots must realize their limitations.

There is a point of confusion in this area that gets people in trouble: the mistaken idea that equipment makes up for lack of ability. A pilot can have an airplane with all the trinkets, bright and new, up to par and working well, but if that pilot doesn't know how to use them, and how to go back and manage with just basic instruments should the fancy stuff fail, that pilot is worse off than a pilot with nothing more than engine instruments and a compass. Remember, despite the most sophisticated equipment available, big airplane or small, it is not possible to fly all the weather conditions Mother Nature can create!

What all this means is that pilots must know and fly by limitations. All pilots, even 40,000-hour airline pilots, have them.

What we are going to do in this book is talk about flying the weather. We will talk about weather in the meteorological sense and then about how we approach the weather problem and what to do about it. We want to talk about pilots' emotions and thoughts when they find themselves say, in the middle of a thunderstorm.

There are things to talk about such as instrument flying techniques and how to test yourself to get an idea of your limitations. There are matters of technique, too, for the person who doesn't fly instruments.

One of aviation's greatest fascinations is the weather. When it's bad it consumes our flying thoughts, but we think about it, too, on a sunny, clear day with a light wind and pleasant temperatures, if for no other reason than we must say, "What a beautiful day to fly. I wonder if it will stay this way." Weather has so many facets that we never stop learning about it, and more than 50 years have taught me for certain that I can never definitely say that the situation is pat and I know exactly what will happen.

We learn to respect weather and never to be complacent about it. This is the warm kind of respect we give to a beloved adversary, for, after all, weather gives us many things: green grass, a blue sky with fluffy white clouds, and the rush of a summer storm that thrills and excites us. It gives the cool, soft kiss of gently falling snow, and the beauty—next day when the front has passed—of sparkling sunshine and crisp, invigorating air.

Like a good friend, weather rarely bores us; it supplies constant variety for our lives. How dull it would be, on the ground or in the air, if we never had to ask, "What's the weather going to be?"

2

A Little Theory for Weather Flying

Most books on weather start out by saying that air is made up of 21 percent oxygen, 78 percent nitrogen, and 1 percent other gases. This isn't that kind of book, and instead we might say that air is made up of wind, turbulence, clouds, precipitation, some fog, and a lot of nice, clear days. But we cannot escape all the theory and must talk about some of it.

You will find lots of things said more than once in this book. This is because many weather factors apply to more than one weather condition and will automatically be repeated as different types of weather and flying are discussed. And lots of points are purposely repeated to be extra sure it's understood they are important.

Weather is complicated. A deep study of the science shows many complex factors, but when you boil it all down, the keys are temperature and moisture, visible (precipitation or cloud) or invisible (vapor). Basically, there is always a certain amount of water vapor in the air, and when air is cooled, this water vapor is squeezed out and made visible. Much of the study of meteorology revolves around the ways water vapor can bother us, and what processes there are to cool the air enough to make that water vapor visible.

✛ That Important Dewpoint

There are a couple of items that pilots should know. One is dewpoint. Most of us know it as the temperature at which condensation begins.

5

If the temperature is 15 degrees C[1] and the dewpoint is 13 degrees C, we have only to cool the air 2 degrees for the moisture to come out where we can see it...and if it's fog, that's all we can see. Dewpoint is handy for a pilot. It's on the weather sequence reports, and it is a simple matter to look at the temperature/dewpoint pair and decide what the chances are for them getting together. That is a matter of deduction and common sense. If night is approaching, the temperature is going down; if there's a body of water with a wind blowing from it toward the land, the dewpoint is going to go up, giving the same effect as lowering temperature. If it's early morning, it's obvious that the sun will come up and heat the air, separating the temperature and dewpoint.

A little warning, however: Sometimes fog doesn't form until the sun comes up. An airport may have the same dewpoint and temperature all during a still night and yet not fog in. Then, just as the sun is coming up and we think everything will be okay, the airport goes zero-zero. The reason for this phenomenon is that formation of fog requires some turbulence. I know that fog feels still and one would bet there wasn't any turbulence, but there is enough to mix the air and give the fog depth; if there were no turbulence, with the dewpoint and temperature the same, you would only get dew on the grass (hence, the name dewpoint), and you would not get fog. As the sun rises, however, on the rare occasion the air is dead still, there is a delicate period when the air begins to stir, creating slight turbulence before its heating is sufficient to raise the temperature. With this sensitive setup, fog may form suddenly; it's like the situation of ice not freezing in a bucket of water with the temperature below freezing until the water is disturbed.

Ground fog will generally burn off with heating from the sun. This burn off, however, will take longer in winter, especially at high latitudes, and if there's an overcast above, reducing the sun's heating, the fog may never burn off. While stewing on the ground waiting for ground fog to go away, ask the Flight Service Station (FSS) if they have any pilot reports on top that say if there's a higher overcast or not; if there is, you'll have a long wait for the fog to dissipate.

If rain begins falling from a front, it will raise the dewpoint, and if night is approaching it will lower the temperature; thus, both temperature and dewpoint are working to get together and make it a miserable evening. There are many combinations that bring the two together, and in most cases it doesn't take a scientist to figure them

---

1. Since weather reports (METAR) now use Centigrade, we'll do the same, with an occasional reference to Fahrenheit just for old times' sake.

out. The temperature/dewpoint relationship is a key guide to weather flying, one pilots should always keep right up front and current.

✛ How Air Cools

We should review, also, how air gets cooler. The simplest way is by the sun going down. As we discover at an early age, it gets cooler at night. This type of cooling is called "cooling by radiation." Another way to cool things off is to bring colder air to an area; a front goes by, the wind turns northwest, colder air flows in, and the temperature drops. Cool air also can flow to the land from a body of water. This process is called *advection*, and it can bring in or take away moisture.

The other cooling process is called *adiabatic*. This is simply a physical law that says when air expands, it gets cooler. What makes it expand? Lifting. When wind pushes air up a mountainside, it is lifting the air. It goes up the mountain to a higher altitude, where the atmospheric pressure is lower. The expanding air cools at a rate of 3 degrees C for each 1,000 feet it's lifted (5.4 degrees F). If the air is lifted high enough, it may cool to the dewpoint, and then a cloud forms.

It's worth noting, too, that the opposite happens when air comes down a mountainside. The pressure increases, the air gets warmer, the dewpoint and temperature separate, and clouds or fog dissipate. The entire process can often be observed on a mountain. The windward side has a cloud near the mountaintop, beginning partway up the slope and hugging the mountainside and mountaintop with an eerie white cover. Then on the downward side we see the cloud shred off and disappear, leaving the downwind mountain slope clear; lifting, cooling, and condensing on the upwind side; descending, warming and dissipating on the downwind side.

We can see the same thing sometimes up high when watching a lenticular cloud downwind and away from a mountain where a wave has formed. As the air flows up the front side of the wave, which is the upwind side, a cloud forms; then as the air flows down the other side of the wave, the cloud disappears. It's a fascinating cloud because it doesn't drift, it just sits in one place, forming and dissipating. It is a smooth, curved cloud, as I've observed when playing with waves in a glider, often skimming close over the smooth, domed top of a brilliant white "lennie," as we commonly call lenticular clouds. But beware of flight under a lenticular cloud; that's where the rotor is located and the air is very, very rough—but more on that later. One can watch the downwind edge of the cloud and see the motion as the pieces shred away and disappear. All this, the flow up or down a mountainside or wave, allows us to see the adiabatic process at work, lifting, reducing pressure, and cooling; lowering, increasing pressure, and heating; making a cloud and then destroying it.

If the air being lifted up the mountain is unstable, the cloud does not dissipate; instead, it keeps going upward and may turn into the type of thunderstorm which we would call *orographic*. That is the difference between stable and unstable air: Stable air comes back down when the force lifting it is removed; unstable air, once it has been lifted to the point where clouds form, breaks loose from the lifting force and keeps going up by itself.

Other things raise air and cool it. A cold front pushes warm air up, or air flows up over cold air becoming, of course, a warm front. And the reverse of nighttime cooling is daytime heating, which makes air rise as thermals, the things glider pilots look for to circle in and go up. When this air cools to its dewpoint, we get cumulus clouds, and if the air has a certain moisture content and is unstable, the clouds grow into thunderstorms.

A meteorologist sees all these changes and additions, or subtractions, in a sophisticated way, studying upper-air soundings and weather information from many places and in many forms. A meteorologist is trained to study this mass of information and analyze it quickly. Now computers are doing most of that task, but in some respects not as well. The computer lacks the local knowledge a meteorologist develops by being stationed at one location for a long time and knowing the little quirks of terrain, wind flow around it, and a host of other things the computer doesn't know. Many old-time instructors and FBO (Fixed Base Operators) have developed a good weather understanding of their area, and a talk with one of them about that airport's weather for future reference can be rewarding.

We as individuals don't have all the sophisticated information, and if we did, most of us wouldn't know what to do with it anyway. But we do have weather reports, and we can see. We can relate what we see to the big weather picture. We can ask ourselves simply, is it going to cool off and is there moisture present, coming in, or possibly coming in, to go with the cooling? By doing this it is possible to reduce all the complex weather factors to a simple understanding. What makes a front potent? Warm air being cooled. What makes clouds on a mountain? Warm air being cooled. What makes fog over a seaside airport? Wind bringing in moisture, or cooling the air. What makes fog in the country? Moist air being cooled at night after the sun goes down. What makes low ceilings when it rains? Rain failing into lower air, raising the dewpoint and causing low stratus and fog. We can go on and on, and finally relate any weather that restricts our visibility to temperatures and moisture.

✝ Season and Time of Day

In our thinking of temperature and moisture, we should consider two important points: season and time of day. In the summer, things are more phlegmatic, and the weather is basically good, or it tries to be. In winter it is more violent and moves and changes quickly. But fall and spring are the most difficult times to predict. Air masses haven't decided whether it's winter or summer; temperatures can be colder or warmer than expected and give unexpected bad weather. The nights are not really long, but they are long enough to produce substantial cooling. A spring day can be mild and docile, or it can blow and be wild.

All our weather thinking should be related to the time of day. We must simply ask, is it the full part of the day when it is warm, or are we catching up with the cool night, when temperatures and dewpoints get together?

✝ Terrain

Terrain is an important ingredient in weather. Terrain which rises presents a chance for air to be lifted. Sometimes this rise in terrain—the orographic effect—can be very abrupt and dramatic. A mountain range may suddenly burst upward from flat ground, like the Rocky Mountains as one approaches Denver from the east after flying over the miles and miles of flat land in eastern Colorado. On the other hand, rising terrain can be subtle, like the gradual slope of the land from the Texas coast of the Gulf of Mexico to the higher land in eastern Colorado. This rise sneaks up on us and doesn't clearly display itself, but the silent flow of warm, moist air up this gentle slope can produce widespread fog, or kick off thunderstorms.

Terrain makes bad weather worse. A cold front being pushed up a mountainside is nastier than a front crossing Indiana, where the terrain is flat. Air-mass thunderstorms are kicked off more quickly when wind flows up a mountainside. Fog can form sooner in valleys where cold air collects. But to make things more cheerful, mountains can help clear up weather on the downwind side, where downflow heats the air and dissipates clouds, or keeps ceilings up. This effect often takes the steam out of cold fronts, making them more docile on the downwind side of a mountain range.

Air can lose its moisture on the upwind side of mountains and be dry and clear on the other side. A vivid example of this appears in the Far West, where the Pacific side of the mountain ranges gets a respectable annual rainfall and supports plentiful vegetation, while the eastern side of these same mountains is desert because most of the moisture is wrung out of the air on the western slopes.

To a pilot over Los Angeles, encountering unexpected bad weather, there's a close escape by flying over the mountains to the desert, where the weather is generally good. All this, in its way, is the adiabatic process at work with terrain helping it.

And when we think of terrain, we should not think only of mountains and valleys, but also of wide streams, lakes, and nearby oceans as well. Water and land generally have different temperatures. In winter the land is colder than the sea, in summer the reverse. You can see a demonstration of the temperature difference between land and sea when flying through the intertropical front along the east coast of South America. During the day the land gets hot, hotter than the sea, and great towering thunderstorms are everywhere—except over the sea, where it's cooler. So you fly out to sea in nice, clear air. At night, however, the sea is warmer than the land and more and more showers are found offshore; so you fly over the land for the best ride.

The lee of the Great Lakes in winter has snow and stratus because the wind blows air across the lakes, where it picks up moisture. Then it blows up the slope of the Allegheny Mountains, rising. The result is zero-zero with snow and clouds on the mountains; the clouds are full of ice, and it takes 9,000 to 14,000 feet to get on top.

Cities are a part of terrain-weather thinking. Cities make smoke and the microscopic particles from it are something on which fog forms. Smoke and air pollution make the formation of fog easier, and a wind carrying pollution toward an airport is a setup for poor visibility. That's terrain, human made, but still terrain.

✝ Wind

Another important factor in weather is wind, which plays a major role in a pilot's life. It affects us from the moment we take the airplane out of the hangar until we secure it for the night. Wind tells us how we must handle an airplane on the ground and during takeoff, it tells us how we must think and act while flying close to uneven terrain, it tells us how short we can take off and land, and what up and down drafts we can expect. Wind affects the performance of our airplane. A big jet weighing 290,000 pounds can take off from a certain runway in calm conditions; with a ten-knot headwind it can increase the gross to 300,000 pounds, but with a five-knot tailwind the gross is reduced to 280,000 pounds. The same rules apply for a Cessna 150; only the numbers are different.

The tremendous force of the wind drove this board (measuring 10 feet × 3 inches × 1 inch) through a palm tree in Puerto Rico during a hurricane. (NOAA PHOTO)

But wind is also important in thinking about large scale weather. First, on a weather map we notice, almost automatically, if the isobars are jammed together like tracks in a railroad yard, saying the wind will be strong, or if they are wide apart, indicating that the wind will be lazy.

Then we look at direction. East winds may bring bad weather, west winds sunshine, wind from a sea such as the Gulf of Mexico, brings moisture that can create bad weather. Knowing what the wind is, or catching its changes in velocity or direction, can give good weather clues.

Wind is layered and blows differently aloft than it does on the ground. The wind up high tells a pilot about speed for a trip, and so, the required fuel and reserves. Wind just above the ground, within the first 1,000 feet, tells about shear and its hazards during takeoff and landing. While the wind may be calm on the ground, especially true in valleys and at night, it can be blasting along at high speed only a few hundred feet above the ground, which can be devastating to a climb rate as you suddenly fly into the unexpected tailwind. Part of preflight weather gathering should be a close inspection of the gradient wind, the wind above the surface out of the earth's friction layer. (Coming into use is the expression PBL which means Planetary Boundary Layer, a fancy name for what we've always called the friction layer.)

An important part of wind action is convergence or, more simply, places where winds from opposite directions bang into each other and pile up. The idea of convergence and what happens because of it is difficult to pinpoint, and the actions which it causes are complicated. A convergence area can be very big, like the intertropical front where northeast trade winds run into southeast trades and create an area of large cumulus and thunderstorms; or it can be tiny, where a sea breeze meets inland air and forms a miniature front of no special consequence except for a line of clouds a little way in from a coastline. These are called sea breeze fronts and generally are mild, but just to keep weather's ability to surprise us alive, thunderstorms occasionally will develop along such fronts. Fronts are a demonstration of convergence, and so are low-pressure areas. The important point is that almost any time convergence is present, there will be some sort of weather associated with it because of the process of air being lifted and cooled.

Divergence is the opposite of convergence. Air flows down and away and, going back to the adiabatic process again, heats up and generally gives good weather. A high-pressure area is a large-scale divergence, a mass of sinking air.

The biggest convergence zone: the intertropical front shown in this satellite picture across the northern part of South America and into the Pacific. It's less active over the South Atlantic, but this is temporary as it strengthens and weakens because of activity or time of day. If you fly between the northern and southern hemisphere, you will have to cross this area that has big thunderstorms—tops into the 60,000-feet range, and heavy rains.

This sinking air in a high, and rising in a low, affects flight more than we realize. Filing a flight plan in a heavy 747 from JFK to Paris, London, or Rome, I'd note whether I was going to be climbing through a high or low after takeoff before I'd decide what altitude to file with ATC for crossing Nantucket, a check point about 176 miles from JFK. If I were going to climb in a low, I would know the airplane could reach 33,000 feet because the converging, rising air would help me climb faster. But climbing through a high—with its diverging, settling air—would be slower, and so I'd file for only 29,000 feet at Nantucket, which is about all the airplane could comfortably make by the time I got there.

I'm sure most pilots—especially flying small, lower-powered aircraft cross-country in a fresh high-pressure area—have noticed how the airplane seemed to fly somewhat slower and work harder to keep normal cruising speed. This is worse in mountainous regions, but that's for another reason—waves—and we'll talk about them later.

A good pilot is wind conscious, aware of its direction and velocity, knowing how it smells and feels, sensitive to a warm, humid wind or a crisp, cold one, where it came from, and what kind of weather it will bring. A good pilot awakes in the morning, looks out the window, sees where the surface wind is coming from, looks up at the clouds, checks which way they are drifting, and learns the wind aloft. All through the day the pilot is subconsciously aware of the wind, and if it changes, senses it and contemplates what this may mean. Any weather-wise pilot puts the wind and flying together and visualizes it tumbling over trees or buildings near the approach end of a runway and what that will do to the airplane. Our pilot tries to "see" the downdraft on a sharp mountainside and relates wind to aircraft performance as well as to weather. The good pilot is animal-like in sensitivity to the wind, feeling and understanding its motions by instinct.

✝ Clouds

A pilot literally looks at the weather to see what it's up to. One of the main things observed is clouds. They tell a big story.

There are two cloud types, cumulus and stratus, and all cloud designations are some combination of them. There are three classifications: cirrus, nimbus, and alto. *Cirrus* are high-altitude clouds, and because they occur in high, cold air they are made of ice crystals; but they still follow the cumulus and stratus designations. *Nimbus* is simply a name given to clouds when precipitation starts to come from them—like cumulonimbus and nimbostratus. *Alto* simply designates height; it means a cloud is at medium height, somewhere between 7,000 and 25,000 feet, and again it is used with the basic cloud forms, as altostratus, altocumulus. You never hear "altocirrus" because cirrus by itself is high.

The important part about the two basic cloud types is their action and this, in turn, tells how they were made. Cumulus clouds are bouncy clouds. They were born of instability, born in air that once it starts up, wants to keep on going—for that's all instability is. Stratus clouds are smooth and flat, or almost flat; their air is basically stable.

Heavy precipitation comes from unstable clouds; steady, light rain or drizzle comes from stable clouds. Said another way, ceilings and visibilities will be high enough to land during unstable conditions, except that briefly heavy rain may cause the visibility to be near zero, the runway flooded, and stopping difficult. Precipitation from stable clouds means low ceilings. Light precipitation can bring zero ceiling, or near it, and the condition can be widespread and of long duration.

So fluffy white clouds are cumulus, and flat, layered ones are stratus. To make it more confusing, they can be in combinations, as stratocumulus, for instance, which is a layer of clouds containing some instability. The precipitation from the clouds of slight instability can be light.

The stories clouds tell are varied. Cumulus clouds are generally thought of as pretty, fluffy white things floating in a blue sky. They mean good weather. But they are not all the same. We know that any cumulus-decorated sky will have choppy air underneath the clouds and smooth air on top. If we look at the clouds more closely, we can get an idea of how choppy it will be underneath. If the clouds have a shredded look, like cotton that's been pulled apart, it's probably rough; you are slapped around the sky, and it's a good bet that the surface winds are strong and gusty. When we fly gliders in these conditions, the rising thermals are generally chopped up and difficult to stay in.

If cumulus clouds are bulbous and fat, however, the choppy air will not be so choppy and the up and downdrafts will be better defined. You rise and descend more like a boat in swells at sea. We also look at these fat cumulus clouds with more suspicion because they are the kind that may grow large and turn into thunderstorms.

We can tell without even looking at a weather map, merely from the type of cumulus present, a lot about the synoptic situation. The first type, the shredded kind, are in an air mass that's close behind a low: a front has gone by recently and fresh, vigorous air is flowing into the area. We're in for a few days of good weather. The fat cumulus

Three decks of clouds: At the top are bands of cirrus, probably part of a fast-moving jetstream; the middle layer, close to us, is altostratus left over from a frontal passage of some hours before; down low are cumulus, which have developed in the unstable air coming in behind the front. This is the sort of story clouds can tell us.

clouds say that we are deeper into a high, perhaps on the back side of it, and warmer unstable air is coming in. Somewhere to the west a cold front is probably starting our way.

Stratus clouds tell a different yarn. We may be flying in a mountainous area such as the New England states. There is a high overcast made of altostratus; the visibility is good. Our destination, in the mountains farther south, is reporting 8,000 feet and five miles visibility with light rain—plenty good enough. We know there's a rain area, a warm front approaching, but the forecasts do not make our destination really bad until long after our arrival. We fly on and notice rain on our windshield. The visibility drops some, but there's enough. We are happy, even though it rains a little harder. But then, looking down in a valley, we see a wisp of stratus below, just a little thin glob of cloud floating along by itself. But it should be a red flare signal! Things are happening; enough rain has fallen into the lower air to raise its dew-point, and stratus is forming; stratus is the cloud low ceilings are made of. It's forming faster than the forecasts indicate; the next thing we know our destination will have about a 300-foot ceiling or less. We review our fuel, check the alternate and destination weather, which is going down, and wish we could hurry and get there before it socks in. All this was told to us by a little piece of stratus.

We are flying westward on a summer day, on top in clear air with excellent visibility; below it's hazy and difficult to see. Way west of us there's a cold front, which is forecast to arrive at our destination long after we do. But suddenly our eye catches a different shading in the high sky far ahead. We take off our sunglasses to see it better, but we don't; we put them on and squint a little trying to pick it out. We fly on and look some more. Then we're certain. The western sky holds solid cirrus, white and innocent looking, but it's a sign that says let's check that front; it may be moving faster than we thought, or a pre-frontal line squall may be developing.

These are a few examples of the many things clouds tell us; they are an entire weather story placed in the sky for us to read. We can study for a long time and never know the whole story, but it is profitable and interesting to try.

3

Some Thoughts on Checking Weather

Prior to any flight, a pilot should check weather; even shooting touch and goes VFR at the local airport information about wind changes, precipitation beginning, and other weather events is valuable.

Having weather and forecast knowledge cuts down on alarming surprises. Being inquisitive about weather must be part of our flying character because it creates an awareness that is necessary even when one's log book shows thousands of hours. It comes down to the fact that the moment a person says, "I'm going to learn to fly," that person needs to add, "And I'll learn weather, too." Flying and weather should be thought of as one skill, one art, never separated. Anytime we fly, weather is part of the game, and the pilot, regardless of mission, should know what the wealther is and what it's supposed to do. We must always be aware, on every flight, that safety is our goal and weather the adversary.

✛ It Isn't Easy

Now, where and how do we get this weather information? It isn't always easy, and there are times we take off feeling the necessary weather questions are not clearly defined. Despite all the information available, today's system still leaves the weather decisions and burden firmly on the pilot, a fact that isn't always recognized.

We get weather from a plethora of services: DUAT, the Internet, fax service, and a host of commercial outfits that give weather, prepare flight plans, and even arrange fuel and food. Our at-home TV weather channels are a wonderful source; we can look and study what the general setup is well before our flight, even starting the day before observing the synoptic configuration and watching its changes and trends, but TV weather is a general picture, one to plan from, not to use as detail weather for flight.

The most important weather information, however, still comes from Flight Service Stations (FSS). There is nothing better than a human to talk to, ask questions of, and get the picture we need of what's out there and how it is going to behave. The problem here is that FSS briefers are not generally meteorologists. Many are very wise about weather, and particularly their area, but they are not free, by regulation, to express personal opinion. Some, on an off-the-cuff basis, will do just that and are very good in their knowledge, but basically they are reading the computer, which displays a profusion of information.

It is possible to talk to a meteorologist, but it's a little tricky. There are meteorologists alive and well in the NWS (National Weather Service) and there are NWS forecast centers scattered across the country. You can call them by telephone, but the trick is that they are not allowed—or set up—to give pilot briefings. However, you can talk to a meteorologist at a NWS facility and ask what he or she thinks about the weather and how good the forecast is for where you're going.

So a scenario might be something like this; I'm going from Burlington, Vermont, to Boston, Massachusetts, so I call the BTV (Burlington) FSS and get a briefing which is not only the weather but NOTAMs, SIGMETs[1], and flying information that is necessary. But feeling a bit uncertain about the weather setup after the briefing, I call the NWS at Burlington, talk to a meteorologist, and ask what his or her idea is of Boston's forecast, what the weather will be like across Vermont and New Hampshire, and how confident he or she feels

---

1. NOTAM: Notice to Airmen; things like an Omni being out, runway out of service, etc. Obtained during briefing and when requested.

   SIGMET: Significant Meteorological Information; issued when significant weather may affect aircraft operation.

   Convective SIGMET: Issued when tornadoes, line squalls, and the real tough stuff is possible.

   AIRMET: Airman's Meteorological Information; issued to amend forecasts and cover moderate icing, turbulence, winds over 30 knots, ceilings under 1000 feet, mountain obscuration, and other nasty things. AIRMETS are thought of as applying more to small aircraft without abundant equipment and performance, but still worth listening to for the big boys. An AIRMET may be part of an FSS briefing.

   These all are issued as needed, and you pick them up by warnings on certain OMNI stations, Flight Watch, ATC, and FSS if you talk to them.

about the prognostications. When I hang up, I have a better feel for the weather.

But what we do with all this information remains the same as it has always been. The key to keep in mind is that a preflight briefing is just what it says, before the flight. Once the wheels leave the ground, the ball game changes. Then it becomes the serious task of keeping up with the weather, watching to see if it's doing what the forecast said or not. "Not" is the important word and emphasizes the need for being prepared to handle the situation a "busted" forecast will confront you with; go to alternate, turn around, climb, descend—whatever.

Weather can change very quickly, and in no way does the currency of information before takeoff relieve us of the responsibility of keeping an eye on what's going on, watching for weather's periodic fickle actions.

A very important point about all this "slick" weather information is that it is still weather with all its faults. The clipped, official look that modern communication and computer or fax printouts have given doesn't relieve the pilot one iota from watching how it goes, being cautiously cynical about forecasts and having alternate action well thought out and provided for—very important.

An important part of any weather briefing is the "IF" information; if it doesn't do this, it may do that, kind of information we always got when talking face to face with a meteorologist, leaning over a weather map, absorbing the picture of lows, highs, fronts, and the rest. To show how this "IF" information works, read this forecast issued:

SYNOPSIS VALID UNTIL 270400

AT 1OZ AREA OF LOW PRES CNTRD OVER THE CAROLINA CSTL WTRS WILL MOVE NWD DURING PD. BY 00Z LOW CNTR XPCTD TO BE LCTD IN THE NEW ENG CSTL WTRS. RMNDR PD LOW WILL MOVE NEWD.

In addition to the synoptic, note the forecast for New Jersey:

NJ

CIGS BLO 10 OVC VSBYS BLO 3SW. CLD TOPS ARND 160. ARND 17Z BCMG 15-25 OVC VSBYS 3-5SW-. OTLK...MVFR CIG.

So our grasp of this situation is that a flight into the New Jersey-New York metropolitan area will have low clouds, but, as we learned from the NJ forecast, quite flyable after 17Z, because the real nasty stuff will be tracking off to the northeast.

Gradual improvement is what it's saying, that the low will move off. However, lows tracking along the coast often have a sneaky way of sliding north, up the Hudson River Valley, slowing down—even stalling—leaving the weather poor and marginal much longer than forecast.

Fronts, lows, highs, and other weather can move differently than forecast most anywhere on the globe, and pilots need be alert to this fact. But what do we do about it?

☦ You Are the Meteorologist

Chances were, in the good old days, the meteorologist you were talking to would tell you how confident he or she was or was not about the movement of that low. So getting ready for takeoff, you not only knew what the weather was likely to be, but also that a suspicious eye should be kept on the development to be sure the low didn't change its mind. In our present age, however, the data flicking on the computer screen has no "if" information, so you have to assume the role—be that meteorologist.

Today the pilot should be weather-wise enough to know when to be wary of a setup and that it is necessary to keep an eye on it. In the latest International Aerodrome Forecast (TAF), now being used in the United States, there's a hint of the forecaster's confidence when you see a statement such as, "PROB40 2022". Which, de-coded, says, there is a 40 percent chance between 20:00 and 22:00 hours of something possibly changing—for example, snow showers, visibility lowering, and so on. A percentage like that in the forecast should make us a little more wary because it admits that a variable exists that they are not dead certain about.

To be a good and safe pilot, you should study meteorology. There are some excellent books that go into the science and make it painless and interesting: *Aviation Weather* by Peter F. Lester and *Weather Maps* by Peter R. Chaston, for example. These and more are listed in the Suggested Reading section.

Obviously, few of us will be meteorologist enough to know all the situations, but we can protect ourselves two ways: one, never be smug about a forecast, and two, keep up with the weather as we fly by periodically getting current airport weather and then comparing it with what the forecast promised. If there is a difference—ceilings and/or visibilities lower, or not improving, if anything isn't as forecast—then it's time to check for more recent actual reports or specials, get new forecasts, and ask if any SIGMETS, AIRMETS, or

Convective SIGMETS have come out. When things begin to look poorly and are not working out as expected, it is also a time to review and recompute our fuel supply and think which way we'll run to stay out of trouble. This is basic and important and becomes more and more so as weather information and briefing increasingly is automated and impersonal.

Periodically there are changes made in the codes and formats of weather information, such as occurred in 1996 when METAR and TAF were introduced and our Fahrenheit world became Centigrade. This change, to conform with the international standards, wasn't a bad idea, considering how much more flying is being done across borders. We may not like change, but it happens, and after our initial griping about it we quickly become accustomed to the new. But it does tell us that it is necessary during our flying life to keep alert for weather dissemination changes and what's new and different.

In the past few years, we have seen the introduction of an Automated Surface Observation System (ASOS) that looks at airport weather, puts it through a computer, and broadcasts it by synthesized voice; there is also better radar information to show thunderstorms and wind shear; and there is other gadgetry of technology. But even with all this, it will still be the pilot's responsibility to get the weather, interpret it, decide what to do about it, and be suspicious of its fickleness. If pilots don't maintain this authority there will be a degeneration of command, and it's not impossible to visualize a ground-based government "dispatcher" telling you if, how, and when to fly! And, of course, a pilot must always be aware that technology can hiccup and miss the weather, tell it wrong, or forget to put it out at all.

There are various ways to keep up with weather as we fly, and more on that later, but there are a few methods worth mentioning now—even if we repeat them later on

1. Ask and listen on Flight Watch—122.0 MHz (EFAS, Enroute Flight Advisory Service).

2. As you fly, listen to ASOS for the current weather at airports along your route and note if they are different from forecast. If so, it's wake-up time to learn what's happening. ASOS is updated continuously, but it has some faults. The visibility measurement is over a very short distance and sometimes confuses haze, ice crystals, and snow. The report never tells you what the surrounding weather is (fog bank north, clouds on hills, etc.), nor does it warn you of thunderstorms nearby or icing. So use it with caution, but if it shows something different

from forecast, get busy on the radio and talk to someone that can supply weather information: Fight Watch or FSS.

There are good points about ASOS, and one of the biggest is that there are far more airports now reporting weather than ever before—via ASOS; look them up and you'll be surprised at even the small airports with ASOS. While parts of ASOS are subject to error, key parts aren't; altimeter setting (pressure), wind, temperature, and dewpoint are accurate. Watching these can reveal weather changes that should spur us into looking deeper into what's happening. ASOS will improve with time, as it's probably here to stay.

3. There are many continuous automatic transcribed weather broadcasts available. Look up where and what frequencies before flight so it will be easy to do when in flight.

There are all kinds of weather information available, and a little study and research will reveal them, but there's always a sure-fire way to obtain the latest information and that is to ask! Call Fight Service Stations for detail weather, and ATC (Air Traffic Control) for SIGMET or AIRMET updates.

4. The most important way of gathering the latest information is to look through the windshield at the signs of weather: clouds, sky, precipitation, ice, thunderstorms, wind on the surface, and anything else in the atmosphere.

In the future we probably will be provided with better weather data, and eventually current conditions will be uplinked and printed out in the cockpit, many airlines have it now. But forecasting will never be 100 percent accurate, so the final weather responsibility always will be yours!

✝ You Are the Captain!

As I've tried to make very clear, it is the pilot's responsibility to get the weather, analyze it, and act upon it. It's more difficult for lower-time pilots to make a decision after looking at the weather briefing material, but regardless, the final decision to go or not must be theirs. Friends who are FSS specialists tell me that pilots frequently ask, "Should I go or not?" This raises the hackles on experienced pilots who have spent a lifetime protecting the command authority of pilots; take that away and you might as well give up flying. It's the sacred part because the pilot finally is responsible—attend an accident hearing and see how true this is as the blame is placed on the pilot, as it is about 80 percent of the time. Aside from that, no one knows how a pilot feels, not just the physical feeling for that day, but how the pilot feels about her or his experience level, ability, the aircraft, and its equipment.

It is appalling to think of a pilot asking a briefer whether or not to fly. If a pilot cannot make that decision and has to ask someone else to make it, then the decision has already been made because if there is that much doubt and uncertainty, then the pilot should stay on the ground; the uncertainty *is* the decision!

4

How to Check Weather

The flight begins when we start to think about and investigate the weather, look over the data, and scrutinize all that information available from the multitude of sources we have already talked about.

What do we check and how, before, during, and after flight?

1. The big picture—the synoptic. Start this a day or so before by watching the TV weather channel.

2. The forecasts—en route and for airports en route and destination.

3. The latest reports, and then look back at past reports—what has it been doing and what is it doing compared to the forecast?

4. Keep alert to the situation in flight by obtaining current reports—what's actually happening, forecast revisions, SIGMETS, AIRMETS, and what you see. Listen to others and be alert to their having problems and what weather information they are asking for. Listen to ASOS for airports you are flying over or near and see if they are conforming to the forecasts, remembering ASOS limitations.

5. Manage the flight with the information and be ready to take alternate action.

6. For learning purposes, and to build experience, take a moment after landing to inspect what happened in relation to what you thought would happen before takeoff. Talking about the flight with a FSS is worthwhile, and justified, and they may appreciate it also because you can tell them what's actually up there.

☦ The Big Picture

Number one, as we said, is the big picture—the synoptic—which tells where fronts, highs, lows, and general weather are. But often, during a briefing, the synoptic is passed over as we eagerly look toward the "actual" weather. This isn't smart; the synoptic should be thought of with care, as it's the foundation on which we build our weather picture. Where does my flight route pass in relation to the fronts, highs, and lows? How would changes in the picture (speeding up, slowing down, or stalling of weather movement) affect my flight? What kind of weather do these fronts and systems have? Ice, thunderstorms, low ceilings, and what do they mean to me in relation to my equipment and ability? And what about the synoptic's past? Has it been behaving as predicted, or has it been inconsistent?

In studying the synoptic, it is important first to *see* an actual picture. Trying to create a picture by reading the messages given from the system is difficult. Adding to the difficulty is the special language computers use, a language developed for their convenience, not yours! This cryptic jargon is difficult to grasp quickly and use to create a mental picture. It's particularly confusing when the message "defines" the location of a line of thunderstorms, or front, by reference to a series of obscure, coded locations with symbols that are only vaguely familiar at best.

For example, try making a synoptic picture in your mind of the following actual message. It's an older format, but the new isn't any better.

SYNOPSIS VALID UNTIL 271400

UPR TROF OVR ROCKYS WITH SFC LOW OVR NM AND MOIST SELY LOW LVL FLOW OVR FCST AREA. FNTL SYS IN GLFMEX WL MOV NWD WITH LOW CNTR DVLPG OVR CNTRL GLF. WDSPRD LOW CLDS/RAIN CONTG OVR MOST OF AREA WITH EMBDD TSHWRS DVLPG AND SPRDG NEWD THRU TX/LWR MS VLY MNLY AFT 06Z.

While a rough picture can be created, the details are missing and, as many old sayings go, there's nothing like a picture—looking at an actual map. If a map has been seen before the above synoptic information is read, it then can make sense and be useful, for example, where it refers to embedded thundershowers developing and spreading northeastward through Texas/Lower Mississippi Valley mainly after 06Z. In today's world of various computer services such as DUAT, Internet, and fax, our chances of having a picture are very good.

Having the real picture in mind, *seeing* the orientation of the FNTL SYS (frontal system), makes it easier to relate those thunderstorms to an actual geographic location. We would know that the de-velopment of thunderstorms "after 06Z" will be dependent on the movement of the fronts and low. By obtaining actual reports as we fly, keeping tabs on how it's all moving—faster or slower—we can apply that to the 06Z forecast and determine how we stand. Many times I've drawn weather maps—crude and simple though they might have been—as I flew, and I traced the confused messages about fronts and storms on facility charts. Recently I purchased an In-Flight Weather Advisory chart made of plastic you can draw on with a crayon pencil and of a size useful in flight. They can be found in most accessory catalogues.

If we are flying on instruments without radar or a spheric system like Strike Finder or Stormscope across that area, feeling smug be-cause we expect to get home well before 06Z, an earlier development or movement may subdue that smugness and alert us to thunder-storms popping up ahead of schedule—perhaps while we're still up there. (Spheric systems tell thunderstorm activity by lightning discharges—more on this later). Being on instruments without radar or spherics with embedded thunderstorms isn't a happy situation!

✝ No Surprises

With this big picture in mind, plus carefully monitoring its movement, we should never have surprises, get caught in nasty situations, or have to use that timeworn alibi, "The forecast wasn't any good."

✝ Satellites

Satellite pictures of the weather are available, and they can be a big help. But we ought to talk about them a little because they aren't the total answer. Rather, they are a good addition to the many sources of weather information.

First, satellite pictures are not the actual raw photo, but are en-hanced by various processes. This isn't something bad. It actually im-proves the picture, but different methods make the pictures a little different. Infrared is used to give temperatures of the clouds. That's how they tell tops, but sometimes, with snow cover in winter, the in-formation is difficult to interpret.

Satellite pictures, like weather maps, have to be related to time, and the most value comes from the latest picture. When we look at commercial TV coverage and their satellite pictures, the time the

A satellite picture (at left) and a weather map, page 31, for the same time. Note that the satellite picture shows a mass of cloud cover extending north of the east-west cold front, a clue that it isn't moving much or clearing out behind it — it's almost stationary — and that the east flow is bringing in moisture over a wide area.

Also note, on the picture, the cloud cover caused by the weak cold front in Texas, plus the squall line, and through north of the low in Oklahoma.

By studying the weather map against the satellite picture we get an excellent idea of where clouds will be that affect our flying.

The luminous white cloud in central-south Texas probably means thunderstorms—a combination of the weak front there and the squall line which says something is cooking. There aren't many thunderstorms showing, but the time is 12:00Z which is still early in the morning over the map area. We'd be certain to check forecasts for future thunderstorm activity.

The gray scale across the top of the satellite picture shows temperature and thus cloud tops. Dull gray is probably lowish stratus, while the bright white indicates very high tops— a thunderstorm indicator.

31

pictures were made is generally unknown; they should show it. With aviation weather services, however, the time of the picture is given.

✛ What Do They Show?

The pictures show cloud cover and, to some degree, how thick the clouds are, but the pictures do not tell cloud bases or ceilings, and they don't tell the height of cloud tops, although satellite experts say they can tell top heights within 2,000 feet. The pictures show convective activity—thunderstorms—but the Radar Summary charts give a better idea of thunderstorm areas. It should be remembered, however, that all these charts, while a good reference for planning how to avoid an area of thunderstorms, in no way should be used to weave *through* a mass of thunderstorms; this takes radar or spherics.

What are the pictures most useful for? First they confirm what the weather map is saying. The clouds of the satellite picture should corroborate the weather map's structure, and it should be easy to pick out front locations as well as post- and prefrontal cloud masses. If the picture and the map do not match, then it's time to dig deeper and learn what's wrong. The first thing to check is how close to each other they are in time. The weather map and satellite picture may have been produced many hours apart, and this could account for the differences. If this isn't the problem, then we should be suspicious that the weather isn't working as the map and forecast indicated it would. We talk about this concept later, but basically it's a job of checking actual reports against forecasts.

For the VFR pilot, the satellite is especially useful because the pilot can see where there are clouds and where there are not. If the flight is over a long distance, the satellite picture can help the VFR pilot plot a course around all clouds. If the VFR pilot finds it necessary to go through an area where the picture shows clouds, then it's a job of checking actuals to see what those clouds are doing to ceilings and visibilities and—very important—what they are forecast to do.

The frequent satellite pictures available in FSSs give an opportunity to study a series of them and see what's been moving and how over a period of time. With this capability pilots can see if the weather is moving off course or becoming thicker and clobbering the route we plan to fly.

By using the pictures, pilots learn more about them and their techniques. Asking questions of the briefer and meteorologist, if we are lucky enough to see a real one, will improve our knowledge of the satellite picture's usefulness. But it is important to remember that these pictures are not the only weather information available and should not be used alone to make decisions. The Radar Summary chart, in the case of thunderstorms, will give a more precise picture than the satellite's. We still need forecasts, actuals, and all the rest, including weather maps.

✛ Where Else Are Those Maps?

Many newspapers carry weather maps, but compared to TV coverage, newspaper data is old and often abbreviated to give a generalized picture of where it's raining, snowing, cold, or hot, but not with sufficient detail to be useful for a pilot, although some newspapers, like *The New York Times*, do a good job. I know airline pilots who buy *The New York Times* first thing in the day to get an idea of how the weather over their routes is shaping up.

There is another value in newspaper weather maps, if they are not too sketchy, and that's in looking back at yesterday's map to see just how the systems and picture have changed since then. Of course, most of us throw the old maps away, but being weather conscious may develop in us a habit to save a few days' maps for reference. Looking back to see how the big picture has acted in the past also can be done in an FSS or weather office, and it's a good idea.

There are other ways old maps can be useful, such as when there's nothing else available. Once I stood in a weather office in Buenos Aires looking things over for a flight to Christchurch, New Zealand, over the South Pole. Since that isn't a regular route, the weather information was slim. The only weather map covering the other side of the globe was a day and a half old. We made an educated guess. Nieut Lieurance, the weather man with us, guided our guessing, and he thought a front would go through Christchurch six hours before our arrival. He hit it almost on the nose, and after a 14-hour flight we arrived over Christchurch in sparkling, clear air washed clean by a recent frontal passage. Not the way you like to do business, but much better than nothing. The big picture had told the story.

✛ Get the Picture First

There's lots of data out there to provide a good picture of the general weather. The key point is to get this picture and visualize it *before* calling the FSS or NWS. Then it will be much easier to relate to the data the briefer reads off for a firm picture of what you will encounter.

Point number two is that your TV look should not take the place of a proper briefing! Important point.

✝ On Days Off, Too

With the availability of TV—and the fact that we look at it daily— we have the opportunity to keep up with weather and be conscious of its movement and development even on the days when we are not flying.

A look at a good, clear TV weather map, not one of the jazzed-up "picture" kind, every day will keep the flow and development and how it's been behaving, erratic or reliable, in mind so that on the days we do fly, it will be easy to pick up the action, understand the situation, and be ready for the demands it may put on us. The weather situation when we fly will be easier to cope with if we've kept abreast of the weather on a daily basis; it's all part of weather and flying being one skill. Aside from all that, it's fun.

✝ The Map

What does a map show? First let's realize that its information is old when we look at it, and by the time we use the information it will be older still. A very important point is that weather is always in motion, it moves and changes, and the movement and change are the things that make it necessary for a pilot to keep an eye on the action all the time.

How many hours ahead the forecast is desired is a key factor in how accurate it is or will turn out to be. Will the ceiling be above limits for the next hour? A forecast of that can be pretty accurate. Will it be above limits five hours from now? That's more difficult to call, and the accuracy has moved from near 100 percent to something quite less. Will the ceiling be above limits five *days* hence? Now the accuracy is way down. So it is important to relate any weather information to the time it was forecast, the time of maps and satellite pictures, the time of winds aloft, and the time an actual report was made. Time difference directly relates to how well you believe, and the caution needed, in the use of any weather information. A pilot should, as habit, look first at the day and time before any weather information is considered.

A weather map is like a snapshot. The shutter clicked at 00:00 hours Z time, UTC (Universal Time Coordinated); that was the moment

What isobars mean on a weather map.

THE CLOSER THE ISOBARS THE FASTER THE WIND - CALLED, "A STEEP GRADIENT". PROBABLY TURBULENCE IN AN ISOBAR "PACKED" AREA TOO.

WINDS CLOCKWISE AROUND A HIGH

FLYING FROM LOW TO HIGH DRIFT IS TOWARD LEFT - CORRECT TO RIGHT - VICE VERSA HIGH TO LOW

CHECK DEPARTURE ALTIMETER SETTING AND DESTINATION TOO - IF IT'S HIGHER THAN DEPARTURE DRIFT IS LEFT - CORRECT RIGHT - VICE VERSA IF DESTINATION LOWER

WINDS COUNTER CLOCKWISE AROUND A LOW

WIND DIRECTION FOLLOWS ISOBARS

ISOBARS FAR APART NOT MUCH WIND - LESS THAN 10 KNOTS HERE.

35

everything stopped, but at 00:01 it started up again, the characters moved in their own way once more, and the map you saw at 00:00 hours is no longer valid.

Surface weather maps are important, but at first look we can be confused by all the numbers and meteorological hieroglyphics. Actually, they are mostly for meteorologists. What may be interesting for us in those numbers we will find on the actual weather reports. But there are important things we do want from a surface map: where the fronts, highs, and lows are; how the isobars curve, because they picture the wind flow and tell us where it's coming from, the velocity, and its source region, which in turn tells what it's like—cold, warm, wet, stable, unstable.

A weather map reminds me of a work of art; the sweep of isobars and fronts is graceful and pleasing to the eye, as is so much of nature's work. But there is more than beauty; there's also an important story that tells where the air is coming from and what it's like— cold, hot, wet, dry—which in turn tells us about the weather.

When we see a deep low in the vicinity of British Columbia, with isobars on the westerly and northerly sides that sweep generally in a southeast direction, packed closely together, we know immediately that cold, wet air is being swept into the area, and we know it's pushed by high-speed winds because the close, packed isobars tell us the wind is howling. There will be lots of wet weather in the area: low clouds and rain with big snow in the mountains, and wind, lots of wind. In the winter in New England, a low off the Maine coast will sweep inland with easterly winds and blizzardlike conditions.

Another day, another map. A northwest flow is over the midwest, with the isobars coming from continental Canada, and we know the weather will be excellent with fair-weather cumulus and good visibility.

A flow from the south and southeast in the center of the United States means hot, humid air is flowing from the Gulf of Mexico and it creates thunderstorms. If a cold flow of north winds is headed toward the Gulf, then the two air masses will meet as a cold front and that will make for very nasty thunderstorms and even tornadoes. This setup is unique to the United States—the cold winds that can flow southward from arctic regions without obstructions such as a mountain range, and the warm, wet winds from the Gulf of Mexico, likewise flowing unobstructed, toward the north. They meet like two warriors and the wild mixing of these contrasting air masses makes our midwest weather famous for its violence and changes. I flew around the world looking for bad weather while doing weather research and never found any worse than in the midwest of the United States, There's not a geographic air-mass flow set up quite like it.

So in that first look at the weather map, we should notice the isobar trajectory and learn from where it brings air and what that area is like—cold, hot, dry, or wet—and how large an area it has crossed that will modify it.

The illustration on page 35 shows some features of a weather map in relation to the isobars. Notice the flow on the east side of the low. The air is coming from the Gulf region and will put warm, humid, sticky air over the eastern part of the country, with thunderstorms, too. Study how the isobars on various parts of the map show where the air has come from: the Gulf, warm and humid; central Canada—called Polar Continental—cold and unstable, bouncy, but good visibility. When inspecting any weather map, be extra wary when the wind-flow pattern brings the air from ocean areas or large bodies of water—West Coast, East Coast, Gulf of Mexico, Great Lakes—and be watchful if you're flying abroad, especially Europe, which has maritime surroundings. Oceans, large bodies of water, and, in a microsense, even rivers bring moisture. And moisture makes weather. (The Seine River wraps around the east end of Orly Field at Paris, France. Early-morning fog rising from the river often makes the east end of Runway 26 zero-zero, with the west end of the airport CAVU.)

WET, COLD
LOW CEILINGS,
Rain, snow, fog

WARM & WET

Thunderstorms,
LOW DEVELOPMENT FOG
Rain in winter

COLD, DRY
CAVU

DRY, HOT, CLEAR

COLD, WET
LOW CEILINGS,
Fog, rain, snow
in season

WET, WARM, FOG
Rain in winter season

Where the heat, cold, wet or dry air comes from to make our weather.

Isobars connect points of equal barometric pressure, but they also show the wind direction, since the wind parallels the isobars to make a picture of the wind direction pattern. The closeness of the isobars tells the wind velocity; if they are tight—close—it'll be a windy day.

(If the flight path is toward a lower pressure, drift will be right, requiring a left correction. Toward a higher pressure, drift will be left, requiring a right correction. This can be picked up with altimeter settings; if the setting ahead is higher, then you'll be drifting left, and the opposite toward a lower altimeter setting.)

So it's the winds, along the isobars, that bring air masses to us. Arguments about what causes the wind and how all this movement

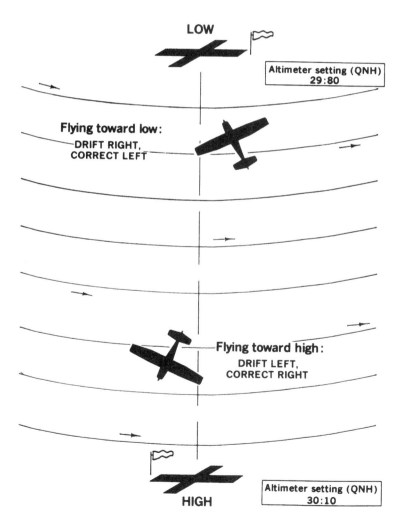

gets started and pushed really don't matter to pilots; we simply know that the wind-isobar pattern is transporting air and it will mix, heat, cool, climb, or get pushed to manufacture the weather.

What else do we look at when we see a weather map? There are two basic items, the pressure systems and the fronts. We look at them and visualize their movements and possible changes in relation to our flight path.

Unfortunately, sometimes we overlook the pressure systems to study the fronts. We are apt to forget that a high-pressure area, sometimes called a ridge, is also a system and not just a place between two lows. A high can often put out a lot of bad weather, and we ought to look at it with that in mind.

Where the lows are is also important. The fronts are a part of them and move or trail along as the low does. It is important to realize how far we will be from the center of the low. A long cold front coming out of a low in Canada and trailing back to the southwest has different weather in different places along its length. The weather at Burlington, Vermont, will be different from the weather at Harrisburg, Pennsylvania, and that will be different from the weather at Greensboro, North Carolina. Each front has its own character, too, and the weather is always different along them. There can be a lot of weather along a front, or none at all—such as a dry front. Sometimes the weather can be ahead of a front or behind it, all of which means you have to study the big picture plus each front's characteristics on sequential weather reports.

Weather is more intense near the low's center. In winter the ice is heaviest, the cloud masses thicker, more confusing, and more difficult to top or fly between; in summer the thunderstorms are wilder. If the flight path will be near a low center, we can be assured that things will be "interesting."

The frontal systems of a low are well known: a warm front, which generally moves south to north; a cold front, which moves northwest to southeast; and, as the low gets older, an occluded front which rotates backward around the low. If the low moves, so do the fronts. If it keeps moving, its movement is easy to forecast, and a pilot will notice that things are working out as advertised. There are stationary fronts, too, which are what the name says—fronts that don't move. They bring messy weather of fog and reduced visibility and, in summer, thunderstorms on a sporadic basis. These storms are difficult to see because of haze, fog, and cloud layers. Stationary fronts aren't violent, like fast-moving cold fronts, but they can cause problems because of low visibilities, especially at night and early and late in the day. The thunderstorms, once formed, can be tough. Stationary fronts make a meteorologist's day trying because it's very difficult to say when the front will move, or whether it will just sit there.

✝ Watch the Slow Lows

Lows that slow down are the nasties that really ruin a forecast. There isn't anything more difficult to figure out than a stalled front. If a cold front was supposed to go right through the East Coast but slows down and stops instead in the New York area, things go to pieces. Instead of clearing, the skies remain cloudy; wind hangs limply around the southerly quadrant, getting over to the southeast perhaps; and fog and low ceilings prevail. With such a situation, a kink may develop on the stalled front, a wave may form, and a new low-pressure area may move up along the coast, following the stalled front's line and putting out a lot of weather.

What this means is that if a forecast calls for frontal passage at 20:00 hours, but the front hasn't gone through by 21:00, it's time to be suspicious. Actually, to prevent surprise, a good weather watcher will check a front's movement as it crosses the country, noting if it passes other stations on schedule. This way a slowing or accelerating tendency can be discovered in advance. Which proves the importance of watching past weather to see how things have been acting. Sometimes past weather reports can be as important as current ones.

Again, wind is important. If it doesn't shift as expected, or its velocity changes, it's a warning sign that something different is happening. Winds picking up generally mean things are beginning to move, probably getting wilder; the rain or snow will be heavier and the air more turbulent. The good part is that the front is on the move and finally will pass, but first there is the battle as it engulfs us.

If the winds slow down in our frontal system, we can worry about very low ceilings and visibilities—perhaps lower than one can land in at this stage of weather-flying art, and the low ceilings and visibilities can extend over a wide area. Depressingly, the slack winds are also a clue that bad conditions will prevail for a lengthy time.

✝ The Wind Speed Tells a Story

The difference between high winds and low winds and the resulting weather is experienced by airline pilots leaving the United States for northern Europe in the winter, a time when Europe has much fog. If winds are slack and there is little pressure movement, pilots worry and take lots of fuel; but if a low-pressure area is approaching the west coast of France, and Paris is forecasting 300 feet with rain and gusty winds, pilots are quite happy and fly off without much concern; there will be enough ceiling and visibility to get in.

We become extra wary in slack winds and bad weather if we have one of the following: our airport is near any body of water such as rivers, lakes, oceans, swamps, etc.; flight late in the day toward darkness; flight toward mountainous areas; flight to a place where there is a lot of moisture on the ground from snow cover in relatively high temperatures, which creates a surefire fog condition; ground soaked by previous rain; and flight toward cities and heavy industrial areas.

✝ Highs Are Not Always Nice

As we visualize a low-pressure area, we often have a dark and foreboding feeling; thinking of a high brings sweetness and light and a feeling of well-being. As we said before, it's not necessarily so. Highs contain, on occasion, fog, ice, thunderstorms, strong winds, turbulence, low clouds, and things we may not worry about directly but should, such as high temperatures that affect our performance or, in winter, very low temperatures that make an altimeter indicate higher than it should. A high has all of these things, and many of them depend on where a low ends and a high begins.

Our low ceilings and poor visibility in snow over the Allegheny Mountains come with a northwest wind, which we think of as the front of a high, although you might call it the back of a low. The unstable air that a high brings to western Pennsylvania will build a cloud deck in winter that often extends as far west as St. Louis. On the mountaintops to the east, the ceiling and visibility will be zero; further west, where the air is older and modified, the tops will be lower and the bases higher. The instrument pilot will have a delightful trip on top, despite battling some degree of ice getting up there and then getting down again; VFR, it's hazardous and across the mountains impossible. Some of our wildest turbulence can be found in a high where strong westerly winds flow over a mountain range and cause standing waves.

Air-mass thunderstorms occur in highs. Most of them are scattered and easy to detour around, but a single thunderstorm becomes a major problem if it's sitting right over the destination airport. To make it confusing, air-mass thunderstorms occasionally line up and, for a while, give the appearance of a front.

The location where the back of a low and the front of a high meet is an important place. We tend to think that when the cold front has passed, the low is gone and we're now in a high, and that means fair weather. Well, it isn't always so, because the northwesterly flow on the front side of a vigorous high may be pumping in wet, cold, unstable air that's often a continuation of the low and its messy weather.

Highs, like lows, move, and when we see a high on a weather map that gives good weather, it's worth visualizing that it is in motion and will depart, and then what is behind it, such as a new low, will be moving in to take its place.

High-pressure areas are often fog generators. The quiet, clear air near the high's center will cool at night, by radiation, to its dewpoint; low land and valley areas may well be fogged in from early evening until after sunrise. How early, how late, and how much depend on the season of the year and how long the high has lingered over the area. Be especially wary of snow-covered ground with air temperatures below, but not too far from 0 degrees C. The air above the snow becomes saturated, and as the air cools with darkness, the dewpoint is easily reached and dense fog forms to stay socked in for a long time.

Northern Europe is a great fog generator at certain times; in the fall a high settles over it and often stays for many days (it's one reason that touring in Europe is best in late September and October—no rain). But the flat high brings nighttime fog, and as the days become shorter, the fog burns off later and later in the day. In October fog forms in the predawn morning and burns off by 9 A.M. or so, but by December it forms near midnight and may not burn off until noon. Some days it never burns off, and I once sat in Paris for five December days waiting for takeoff minimums! No particular hardship.

So while we're sitting in the middle of a tranquil high, airplanes are having a difficult time landing and may divert to various airports in Europe. Flying 747s to Paris, I've made many an approach to 400 meters visibility in this fall-winter high. A good aspect of it, however, is that in such flat highs the winds are calm or light; otherwise there wouldn't be fog. An instrument approach is easy to fly because there isn't any drift or shear.

The back side of a high—or we could call it the front of a low— has southerly wind flow that kills off the probability of fog because the air is warmer, doesn't cool off as much at night, and is moving. This inflow of warmer, more moist air is where the afternoon air-mass thunderstorms develop. And this air, moving from the south, runs up over the back side of the colder dome of high pressure and begins the process of the warm front of the next low-pressure area. Milky, high cirrus clouds that dull sunshine tell us, in an upward glance, that this process has begun.

So high-pressure areas need serious attention, too; what's the wind, how old is the high, are we flying in its front, center, or back? It's all important.

✢ Look up

But we are apt to think only of the surface map when, really, that isn't where we fly. Fortunately there are maps drawn for where we do fly, in the sky above. These are often overlooked and they should not be.

They show wind patterns, velocities, and temperatures at different flight levels. They aren't listed as altitude maps, but as pressure-level charts expressed in millibars (mbs). Sounds complicated, but it isn't, and all we need to think about is these millibar levels as altitudes. Various-level charts are drawn, starting at 850 mbs, which is roughly 5,000 feet. So if 5,000 feet is where we fly, then a study of the 850-mb chart will give a more realistic picture of what the winds are and what's going on up there.

These pressure charts are made for:

850 mbs	5,000 feet
700	10,000
500	18,000
300	30,000
250	35,000
200	39,000
150	45,000

There are higher charts for 100 mb and 50 mb, but we don't get them in most FSSs or weather offices, and we don't fly up there much anyway; that's Concorde country!

The altitudes are approximate because the maps actually show at what level, in meters, we find 850 mbs, or whatever. So the charts, if viewed sideways, would undulate as the pressure levels do, reflecting lows and highs.

All we're doing in this little exercise is to show how the millibar-level charts relate to altitude and to say that if we're flying near 10,000 feet, we should study the 700-mb chart and others for other altitudes.

The locations of highs and lows are different aloft than on the surface. At high altitudes, one may not even see a low because it's all down underneath, and the 200-mb chart, 39,000 feet, may have the isobars in a straight line with the low nowhere in sight. If it does show up at the 200 mb level, however, you can be sure it's a lulu!

The sequence of charts starting on this page shows the changes between the surface and 300 mbs. The surface chart has a deep low on the Canadian border west of Lake Superior. At 700 mb, 10,000 feet, we see the strong wind gradient on the south and southeast side. At 500 mb, 18,000 feet, the low isn't as intense and the isobars start to become more west-easterly. At 300 mbs, 30,000 feet, the intense, tight circulation has smoothed out and started to join the zonal west-east flow. Most of the action is down low, and the 30,000-foot-plus jet airplanes will have little weather beyond ice-crystal clouds and probably some light turbulence. The charts illustrated here were drawn from actual charts for the same day and time.

According to the 700-mb chart, if we were flying from Seattle toward San Diego, we'd have a constant left drift because we would be

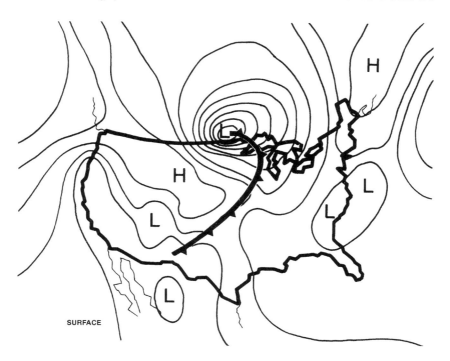

Series of four mb charts—same day, same time, different levels. FSSs and weather offices have them.

700 mb analysis height/temperature (about 10,000 feet).

500 mb analysis height/temperature (about 18,000 feet).

300 mb analysis height/temperature (about 30,000 feet).

flying toward a high and would require a plus drift correction, although not much because the winds are light along that route. We can judge the wind velocity by the isobars or the barbed arrows on the chart.

What we're trying to get across is that those upper-level charts are very important, and not enough use is made of them. I always consider which chart to concentrate on in relation to the airplane I'm flying: If it's a small, single-engine I look at the 850- and 700-mb charts, the charts up to 10,000 feet. For a turboprop, the 500 mb is the interesting one because I'll fly in the vicinity of 18,000 feet, although some of these airplanes are getting up above 30,000 feet. In a pure jet I look at 300 mbs and up. Of course, I take a glance at them all to see which way the lows are leaning, how deep they are, and what's going on at other levels. Again, these charts are important because that's where we're flying and we should know what's going on up there.

5

Weather Information

When we have the synoptic and map well in mind—and remember that the upper air charts are part of the synoptic—the next step is to study the weather in detail to know what's actually going on: actuals, upper-level charts, radar summary charts, pilot reports, and forecasts—there is more, too.

✢ Always Learning Where and How

As I said previously, there is constant change in the weather dissemination process, just as navigation aids change. So it is necessary to review, frequently, what changes have been made, what the latest methods and codes are, and where to find them. The responsibility for keeping current in all divisions of the flying world cannot be taken lightly. Those who have only a shallow knowledge will eventually get in trouble.

There are publications important to this end. The *Airman's Information Manual* (AIM) in the United States will give a picture of what's available on the ground and in the air. I find that talking to FSS people and asking them what services they provide gives me an excellent education and update. Of course, that has to be done when the FSS specialists have some free time—probably during a CAVU day. Users can surf the Internet for weather information and partake of the computer and fax services available. Reading aviation magazines for a general idea of coming technology and the newest available technology is not only interesting, but almost a necessity.

✛ An Extra

The National Oceanic and Atmospheric Administration (NOAA) has a network of stations that continually broadcast local weather on frequencies from 162.40 to 162.55 megahertz (MHz). These are not specifically aviation oriented, but they give a good picture of the weather over an area of roughly 100 miles. They are changed three times a day, except if there is an emergency or sudden adverse weather developments, when they're changed immediately. There are about 375 stations around the United States, so most of the country is covered and about 90 percent of the nation's population is within range. Inexpensive special receivers can be purchased for this reception at most radio stores, some with an alarm which warns of any imminent dangerous weather. We have one here at Sugarbush, Vermont, airport and it has been useful. One warm summer afternoon, the alarm alerted us to an approaching line squall. There was a mad rush to tie down gliders and aircraft. The warning probably prevented aircraft damage as the wild wind and rain rushed in.

✛ No One Said
It Was Easy

It's difficult to sort out the mass of acronyms, as well as the different and overlapping services. Vague references to where you hear weather and on what frequencies compounds the muddle.

Regardless of all these problems, the need for weather information is still there, and when you climb in an airplane and prepare to fly, it isn't any different from the way it was when things were simple. It is important to mentally set aside the "modern" methods of communication—the confusion of frequencies, telephones, computers, and mass of data—to clear your mind and focus on the gut things that count: fog, ice, thunderstorms, ceilings, visibilities, wind, turbulence—the lot—no matter how it's packaged. You must dig for information and ask questions until your mind has a clear picture of what's out there and how it may or may not act.

How do we sort it out? As we've mentioned, FSSs give briefings of present weather, forecasts, a condensed synoptic, winds aloft, and any weather warnings in force. The National Weather Service offices do, too. But, as I also said, the weather given to the pilot presents a picture of weather straight ahead, down the course line. It tells little of the action off course where fronts or other kinds of weather may be in motion to eventually create problems along your course.

✛ A New Way

In recent years, private meteorological services have become available. You pay money and get a service. There are many reasons these services have come into being. One is that FSSs are busy and, at times, difficult to contact. The new super FSSs that service a wide area are even more difficult to get on the telephone. One corporate operator I talked with said the private weather service can complete the briefing and file a flight plan in the time it normally takes just to get an FSS briefer on the phone! The private services also make up flight plans fitted to your airplane and, in some cases, file them for you. Some even get all your international paperwork organized and obtain permits. They also have toll-free telephone numbers that plug you right into a meteorologist you can talk with about the weather to allay your doubts—or at least explain things.

The idea of these services is excellent and can give the pilot what's really needed in weather. Unfortunately, all such services are not really meteorologically orientated and, in effect, are simply getting the information from government services and putting it in their computer bank, organizing it a bit, making a flight plan from it, and shipping this off to the user with no weather interpretation or explanation. They haven't done anything you couldn't do yourself if you want to go in an FSS or an NWS office and dig it out.

If you are going to use one of these services, the important key to check is that the service you pick has meteorologists who interpret the data, massage it, and talk to you about them—like the good-old days. Without that part of it, it's just a streamlined way of getting the data you could get yourself.

Of course, these private services cost money, and the average private airplane owner isn't likely to add the extra costs when the government supplies a weather service.

As we see more and more of the paid services at fixed base operators (FBOs), plus the ability to obtain them privately via telephone, computer modem, or satellite, we wonder if this isn't the way of the future as the government's weather dissemination methods seem to become more impersonal and slow to respond.

But private service, Duat, Internet, FSS, NWS, or what have you, the game is the same. The pilot must make the decision and must ask the right questions to have a proper picture and knowledge of weather to fly by.

✝ Opening Remarks to the FSS

As small a point as it may seem, what you first say to an FSS or NWS briefer is important. The key is not to glide into the office, or pick up the phone, to simply say, "Hi y'all, how's it look to LA?" If you do, many briefers will tell you, just as unceremoniously, "It's VFR" or "It's not very good" or whatever. To get the weather you need, VFR or IFR, it is very important to open the conversation with, "I'm a (airplane type) "N" number__, I can go IFR (or VFR, whatever the case), and I'd like a briefing Sacramento to LA departing about 13:00 local."

Now you've asked for the works, which is what you should ask for and pay attention to. And the briefer knows you're serious and want a good picture of what's out there—synoptic to sigmet.

Doing it off-handed and informally invites an off-handed and informal response and, for those few briefers who like to tell you how to fly, a chance to editorialize rather than detail the weather, which often finds the pilot departing with a sketchy idea of what's really out there. It may sound stuffy, but this is the time to be formal and serious!

Much of the time we have to do the briefing over the telephone, which makes it more difficult. But three things help:

1. Look at TV weather in advance and know the general setup for reference.
2. Don't be afraid to ask the briefer to repeat things.
3. Train yourself to write down weather symbols so you're using a sort of shorthand to copy what's being said.

It is useful to have a small blank map handy on which to draw a picture of what you're given for the synoptic situation. It is a good way to bring the weather picture into focus.

✝ Synoptic Again

So where do we start? With our old friend the synoptic, no matter where or how we get it. If you have a good mental concept of this big picture, you will know if there are fronts on the move and troublesome weather stirring around for your flight. What you are trying to learn is what to be wary of: weather speeding up, slowing, stalling, changing to lower ceilings for landing; thunderstorms or ice developing, or already in the picture, that will seriously harass your flight. So—important point—if your synoptic homework was done well, then you are in a position to ask, after the routine briefing, some searching questions about off-course weather that might worm its way onto your route, making the day difficult. Where is the front that's not scheduled to be over the route until after we've landed? Any chance it's moving faster? Fast enough to clobber the route while we flying over it? How about those thunderstorms forecast for tonight? Might they mature in the late afternoon—about the time we're expecting to be nearing our destination? These are the important questions, and the protective intelligence to ask them comes from what we learn scrutinizing the synoptic in advance.

✛ Look Ahead

Much in tune with the above is the wisdom, when getting a briefing, of asking for future forecasts—forecasts for the next period after your ETA so you can keep an eye and ear out for the possibility of weather moving or developing sooner and giving you trouble. Such inquisitiveness backs up the basic fact that one should always approach weather forecasts with a certain amount of skepticism.

✛ The Real Thing

Now on to the actual weather reports–what is it really doing? These should be studied along the entire route, not just for the airport of destination, but for each reporting point along the way so we'll know what to expect en route. This is especially important for the VFR pilot, and even more important if there is rough terrain—mountains—along the way. The higher terrain will have lower ceilings and visibilities under conditions of poor weather. A scrutiny of the reports will give an idea of what to expect.

There's nothing more sloppy or hazardous than the pilot who simply asks, "Is it VFR?" and then armed with that, and perhaps winds aloft, charges off into the blue, flying from a position of ignorance.

The study of the actual reports for airports along, behind, ahead, and to the side of the route for the latest and past few hours is important for learning how the weather has been acting. To be thorough, a past forecast should also be looked over and correlated with the actuals to see how the weather has been, compared to what it was supposed to have been; this may be difficult to obtain.

When in the air, the weather we're flying in should be related to each hourly report, which should be gathered routinely and religiously while flying. The past reports, compared with the new hourlies, can give a strong insight to weather changing, developing, deteriorating, or, happily, getting better. This constant study, while flying, of the latest actual weather reports cannot be emphasized too strongly. Equally important is the wily pondering of what's changing, which sends signals that not all's well in the sky. The key point here is that the actual reports should be gathered each hour as we fly and compared with the past hour's and the forecast!

Obviously, if we're flying across a big high and it's CAVU all over, we don't have to check weather every hour, but the more inclement the weather, the more we have to be on top of it. It doesn't hurt, either, on that CAVU day, to gather a report or two to know that it's really staying CAVU. There is always the possibility for surprises, and if the weather is that good, there isn't much to do anyway.

It's only natural that our major attention is toward the destination, but there's much to be learned about development along the way, especially if thunderstorms are in the forecast as well as ice and turbulence. So attention needs to be given to reports for stations along the route as well as the airport we're headed for.

Constantly, however, we check the destination's hourly actuals to be certain the weather is holding up and not starting a downhill slide of ceiling and visibility. There's little or no excuse for a pilot to arrive over the destination airport and be "surprised" by below-limit conditions.

Actual weather reports are tremendously important. Because weather forecasting is not exact, and probably never will be, we need the actual reports to know how the forecasts are doing: the weather as promised or something different?

The actual weather report is the point of truth—all the charts, maps, and reports we've seen, or forecasts we've been given, are not *real;* they are only estimates of what will happen, and the radar charts, satellite pictures, and all that are what was, not what is, when you get up there. The METAR is the latest information, but remember, it still may be different when you arrive; we start an instrument approach with the weather reporting 200 feet and half a mile, within our limits, but when we get to 200 feet, the ground doesn't show up, and the ceiling has gone down since we started the approach. Now what are you going to do? What I'm pointing out is that finally the flight will be directed by the pilot in command who, we hope, understands that the real conditions affecting the flight are what is seen and experienced, not what charts, maps, or reports say. Now the pilot is in command, performing the best way possible, using whatever experience and skill is in that pilot's background, to complete a safe flight.

But in weather, gathering these things (the synoptic picture, forecasts, and actual reports) are the key. If this book only got that across, it would be worthwhile.

Many years ago, teletype sequences went west to east and south to north, giving actual weather reports in geographical order. Checking weather, one read a tape of the reports, like the old Stock Market ticker tape. That was logical because weather moves in these directions. You could observe a front moving toward the east as you read each station's weather report in line. The wind would be northwest at Kansas City, but still southwest at Columbia, Missouri, and you knew where the cold front was—somewhere between the two stations. Or one could see a warm front progress from south to north, graphically, because the reporting stations followed the same path. Computers have taken over, and all this has become history.

Today, being briefed, you get the actual reports for stations along the route you'll fly—not beyond or before, or off course. It's as though you're looking down a route with horse blinders on. But what may be beyond the station moving toward it, or a front off to one side, and what was previous weather compared to the forecast? All things a pilot should know. The information is available, and you can ask the FSS briefer for it, then study the setup and its possibilities yourself. This is necessary, and a pilot should have the discipline to do it. The present system will tend to make the lazy pilot take the route forecast and go with it—a good weather pilot will dig deeper.

The pilot must keep digging until satisfied that there is good and sufficient information to tell the weather story. This can be difficult on telephone briefings because the person on the other end of the phone may show signs of impatience as you ask for more information than the normal briefing gives. (This is rare, but happens.) They are busy, but that shouldn't intimidate us.

I was talking to a briefer on the telephone one day about weather from Philadelphia, Pennsylvania, to Burlington, Vermont. He said it was all VFR. But from the TV map I knew there was a front to the west that could move on course if it speeded up. I asked about it. "Oh, that won't bother you. Your route's forecast VFR for your time period."

Not satisfied, I started asking for actual weather reports at stations like Buffalo, Rochester, and Harrisburg in an attempt to learn where the front really was. The briefer became irritated and said rather strongly, "What do you want those stations for? You're not goin' that way. I just told you it's VFR!"

Which demonstrates why people get in trouble with a sketchy picture of the weather. That front was moving faster than forecast and did mess up the course, and I had to go IFR after I started. But what about the VFR pilot new to the game and starting out on that briefing? Would he or she be smart enough to land? Turn around? Run east, or press on, getting lower and lower, with less and less visibility in mountainous country, thinking it ought to be okay because the briefer said it would be VFR? It's the typical setup for a weather accident by pushing VFR too far.

This makes the point, also, that a pilot cannot depend on a briefing being so accurate that the statement "It's VFR" covers everything. Weather science isn't accurate enough to do that, but the tendency of the system is to make it appear as though it is, and this means another adjustment in pilots' thinking; they must not allow the official-looking maps, charts, forecasts, and briefings lull them into feeling that the statement "It'll be VFR" is solidly accurate. In the air, in the weather, in clouds, ice, thunderstorms, fog, or whatever, the authoritative appearance of the weather briefing office quickly fades as a pilot combats the weather and wonders what this is that the briefing didn't divulge—and now what to do about it!

Preflight study and constant updating will be the pilot's defense and aid when there's a sudden discovery that the forecast—and the statement "It'll be VFR"—are inaccurate and the real world is different and tougher!

Of course, a pilot should feel free anytime to pick up the mike and ask for weather. But remember, getting weather information is your job; nobody is going to spoon-feed it to you! It is easily possible to be flying along feeling good about the world, never realizing there was some tough stuff ahead or at destination. This euphoric, and dangerous, situation could exist simply because the pilot did not assume the responsibility of keeping up with what's going on with the weather.

Knowing weather, and keeping up with its changes, pilots should realize that their flight route or destination may have to change with the weather. It is folly to plug along without flexibility of thought and willingness to realize that one may have to fly differently, change route, go IFR, divert, turn back, or even land!

This is a significant point—being flexible. The fixation of plugging toward the destination, even though the weather has turned foul, has killed too many people. One should be ready and willing to toss aside the idea of "getting there" in favor of turning away to a safe-weather airport. Important!

These problems of gathering and studying weather information, I believe, are the genesis of more weather accidents than any other single factor. The system has seriously reduced the pilot's ability and incentive to make weather judgments and be a flying weather person when, as we've said and will say again and again, a safe pilot must be a weather person too!

6

Checking Weather Details

When we look at details, chances are that the first thing we'll glance toward will be the destination weather. What is it, VFR or IFR, and how close is it to the legal limits?

It doesn't take too much soul-searching to judge if weather is within one's ability and equipment. Basically, the flight can either be made VFR or it requires IFR, and marginal weather between the two can cause trouble. If it is marginally VFR, then the pilot had best be prepared to go IFR in case suddenly, en route, it's discovered that IFR has become a necessity. Trying to hang onto VFR when things have deteriorated and IFR weather is becoming more a reality each moment is a proven method of getting into serious trouble!

✝ VFR Not Easy

If it's decided VFR is going to be the way of the flight, then the pilot may have a more difficult decision than the IFR pilot. In many respects the VFRer needs to learn more about the weather setup than the IFR pilot because the VFR pilot must know if the weather will remain VFR en route as well as at the destination. Sometimes this demanding decision is being made by the least experienced pilot! But experienced or not, the pilot should make the decision and not simply depend on a briefer's statement that "It's VFR all the way."

The statement that should get a pilot's attention, and be considered seriously, is the briefer saying, "VFR flight is not recommended." The final decision is still the pilot's, but VFR not recommended should be a warning to dig even deeper for information.

✝ MVFR

The Weather Depiction Chart shows areas where conditions for IFR, VFR, and MVFR (Marginal VFR) flight can be expected. MVFR is also used in forecasts. It's the MVFR portion that should be considered very carefully. It would be more realistic to change the label of MVFR to HVFR, which would stand for Hazardous VFR! Why? Because that's what it is, hazardous! MVFR is defined as "ceiling 1,000 to 3,000 feet and/or visibility three to five miles inclusive." This can be dicey weather for VFR. Poking along, down low, in mountainous areas with such reduced ceilings and/or visibilities is asking for trouble. If precipitation is falling, then it's much worse because three-mile visibility with rain smearing a windshield reduces what you can actually see to far less than three miles.

Imagine. We're nervously sneaking through this murk in mountainous country, maybe a high peak just ahead, or a TV tower that reaches into the sky to a scary height and suddenly appears out of the gray world—a hazardous way of flying.

Ceilings are based on broken clouds to overcast. This means scattered clouds can be floating around in the falling precipitation of an MVFR area below the 1,000- to 3,000-foot ceiling. So as one flies in the rain-reduced three miles, clouds can be encountered almost half the time, because scattered is defined as 0.1 to 0.5 sky cover. Bumping into these wispy pieces of cloud or twisting along trying to miss them can add to the difficulty of struggling to stay VFR. You can also feel nervous when it's raining and lower scattered clouds are encountered because they most probably will increase and become broken to overcast as the moisture of the rain raises dewpoints. Then the ceiling could become much below 1,000 feet—and the visibility lower, too.

✝ MVFR Is Not Static

An additional, and very important point, is that MVFR areas are not static. The weather doesn't stand still very often or very long. Remember, it is always in motion! Poorer weather may be on the increase, or things may be getting better. But either way is tricky because it can get worse sooner, or better later, than expected. The pilot planning to fly into that MVFR area needs to be very aware of the weather prognosis and study it carefully before flight to learn which way conditions for the MVFR area are forecast to change, up or down, and then watch it carefully in flight by gathering weather reports and looking through the windshield to see how the MVFR is behaving. And what you see, or cannot see, ahead is the moment of truth, the real thing, the actual weather—regardless of the forecast.

In VFR there is little time for decision. One cannot keep pushing hoping it will get better. The decision to turn and run or go IFR is now!

One can lay down a basic rule that anytime you are headed into MVFR areas, you should be willing and able suddenly to go IFR. There's no way of knowing, but bets are that MVFR displayed on charts has suckered people into flying areas where they never should have—and caused accidents.

✝ IFR Not to Worry

The IFR pilot doesn't have to worry about clouds en route or terrain because IFR flight is at safe altitudes above it all. A pilot doesn't fear flying in clouds—or shouldn't. But the VFR pilot who hasn't an instrument rating has a very strong concern not to get into clouds or areas of reduced visibility that do not give enough reference to fly the airplane and maintain control, as well as see what's ahead.

As I have said before, and warned that I'd repeat because it is so important, once a pilot looks at the weather and decides it's good enough, the question to be answered is: Will it stay that way? How do we answer that? Back to the big picture to see if weather is approaching in the form of a front or if there is air-mass deterioration; or possibly, and happily, that a front has passed, indicating improvement; or if the airport is near the center of a high, which affords protection on both sides. An exception may be early-morning ground fog, but that will burn off soon after sunrise.

The destination isn't the only area of interest, however. We want to consider alternate airports for both our destination and departure points. Either VFR or IFR, we may want to return and should know if the departure field will be accessible, and if it isn't, where we can go in case of trouble when we are still close to our takeoff point. Regarding airport weather, then, we are interested in destination and alternate, departure, and departure alternate.

If the destination is too bad, then a pilot wants to know when it is going to improve. What do we look at first? The forecasts. The best way of studying them is to check them over a period of time. We check for the current time—that is, what the forecast predicts for this very moment—then for the previous four hours, and finally for the period of our expected arrival and a few hours after that.

✝ Test the Forecast

With all this firmly in mind, we go to the actual sequences and see how the forecasts have been performing. What *was* the weather during the previous four-hour period, compared to what the forecast said it would be? Then the present: What is it actually doing compared to what the forecast says it should be doing? Note how many special observations have been made. Specials show that something is different—changing—and should act as a signal to look closer and be alert to what the changes are. Now we have a feel for the weather, for the kind of weather it is and whether or not it's acting according to plan. If it hasn't been doing as expected, we'd best look again at the big picture and try to decide what's been going on. Are the fronts moving faster or slower, are the lows slowing, has a flow of air from the sea remained instead of turning around? The important point is to give the weather forecasts a test and see how they have been performing. Good performance can indicate an easily forecast situation. A bad forecast generally means a tough setup that has the weather person guessing. We learn to look at forecasts with a degree of confidence based on their past performance.

Once we have seen what the weather has done and what it is now doing, it's only necessary to fill in what it will be doing when we get there.

✝ The Late Weather

There's another point worth consideration: what our airport is forecast to do after we get there. We should know the expected weather for a period of four hours after our arrival. This isn't in case we're late, although that could be important, but rather to give a picture of the trend. If, for example, the airport is forecast clear and unlimited for a long period after our arrival, we feel more relaxed about the chances of its being good when we arrive. If, however, it's forecast to begin a gradual deterioration four hours after our arrival, then we keep an eye on the possibility of this deterioration occurring much earlier than expected. What we are doing is bracketing the expected weather for our arrival; we get the forecast for *before*, and finally, *after* our arrival. What we do is slip the time of arrival that we will really use in between the current weather and the future weather, and see toward which time, now or future, it seems to be developing.

✝ Regulations Aren't the Important Criteria

The things we look at to start with are ceiling and visibility. Are they forecast to be within VFR limits, or will an instrument flight be necessary? This question involves the Federal Aviation Regulations (FARs), which say that you need a certain minimum ceiling and visibility to fly. What these minimums are is determined by regulations, but an important point about legal minimums is that they are a batch of words in a book and aren't to be considered a substitute for good judgment. An airport may be forecast to be better than VFR limits and still not be a place to fly. Suppose it's down in a valley surrounded by mountains that are cloud covered; there just wouldn't be any way of getting in there VFR. The legal minimums may not always be good enough for some situations.

When looking at ceiling reports, remember that the ceiling reported isn't for the scattered clouds, but for the broken and overcast. So if a ceiling is reported as 800 feet with scattered at 200, realize that you may have ground contact at 800 feet, but there will be annoying scud occasionally blocking your visibility much lower. If it's raining, you can bet that the scattered will become broken or overcast and become the 200-foot ceiling later on.

Visibility is an important factor for all pilots, whatever their experience. If ceilings will be at or near minimums and the visibility is good, then we know it won't be difficult to fly, but if visibilities are low, a minimum ceiling makes flying much tougher. If we are poking around mountainous terrain with low ceilings and low visibilities, we are in a difficult situation. A visibility of five miles with a 1,000-foot ceiling doesn't sound bad in Indiana, but that same weather between Winslow and Kingman, Arizona, would be very hazardous and difficult.

The same applies to the low approach. A 200-foot ceiling with a couple of miles' visibility isn't a difficult approach if turbulence and wind are reasonable, but when the visibility drops to half a mile, the approach is much different.

Conversely, we can handle reduced visibility if there is a lot of ceiling so that we can get up high enough to clear all the terrain we cannot see. In practice, however, we prefer visibility to ceiling if we cannot have both.

Whenever we consider visibility, we should note whether or not there is precipitation. Four miles reported by the NWS can be much less when seen through a rain-smeared windshield. Four miles may in reality be only one bleary, wavy mile. Airplane windshields vary in their ability to shed precipitation, and some are really awful. Many aircraft are now being equipped with windshield wipers, which help tremendously. I don't think we realize how much they really do help. Pilots may not think there's any problem because they can see through the windshield in rain; but if a wiper were suddenly turned on, there would be a dramatic increase in what's visible and we'd realize that we had been cheated out of a couple of miles of visibility all the time without wipers. So the airplane windshield and how well you can see through it in rain is an added weather factor. Make this little test; the next time you're flying in rain straining to see ahead, look out a side window that isn't getting rain from straight on and see how much better the visibility is.

☨ Pollution and Visibility

Visibility is a function, also, of smoke and haze coming from industrial areas. Perhaps an airport is reporting reduced visibility, while the general weather conditions are good. The reduced visibility may not be cause for alarm because it could be industrial air pollution drifting in from the city. A look at weather reports from other airports in the area that are not downwind of a city will show the real visibility trend.

With a light west wind, which says basically that the weather ought to be good, LaGuardia airport at New York could have one-mile visibility, when Newark airport, upwind of smoke-producing areas, may have seven miles or more.

An airport's location can mean a lot. If Long Island, New York, has a northeast wind and there's worry about fog and low ceilings, it's a cinch that LaGuardia Field, on the north shore, will have worse conditions than Kennedy on the south shore. The wind has to flow slightly down and over land to reach Kennedy, and this will help to raise the ceiling. What we're saying, again, is that we must include the factor of terrain in our weather evaluation.

If we are going to fly instruments into an airport, we can accept a forecast that calls for weather within legal limits for an instrument approach. But these limits, again, are only legal verbiage and because it's legal doesn't always mean you should go. Legal means regulations with fixed numbers like "200 feet," but numbers cannot cover all the factors in weather flying. Wind and turbulence can affect an approach and may make a 300-foot ceiling seem awfully low. In an extreme example, you might have a legal 300-foot ceiling forecast for Daytona Beach, Florida, during a hurricane! You certainly wouldn't want to go there. To some degree, these elements enter into all weather judgments. If it's going to be necessary to descend through a thick layer of icing and a pilot's experience and anti-ice equipment are limited, the pilot should not be interested in making an approach even if it is legal. There are a lot of common sense factors in judging whether to fly or not, and legality shouldn't always be the deciding one. It is folly to think that as long as there are legal minimums everything is okay; there's more to it than that! Of course, a pilot cannot operate *below* legal minimums unless ready to declare an emergency or pay fines.

There are other times when the legal minimums look too restrictive—for example, a low stratus condition with calm air and little wind. In such a situation a person could probably cut a ceiling below legal minimums without much effort. Generally, however, legal minimums are something not to go below, and there are times when one should remain above them, times when they are too low.

Cheating on minimums isn't clever. It's well worth remembering that minimums are based on many things, and the important ones are terrain and obstructions under the descent path, to the sides of it, and in the missed approach area. Sneaking below minimums means taking the risks of hitting something.

✝ How Do You Feel?

Another serious factor in weather judgment—often overlooked—is how a pilot feels. Some days we feel good and very tigerish; we've had good rest, our physical condition is tip-top, and we feel able. At other times we may be tired or have a cold that lowers our efficiency. We may be unhappy because of anything from a bad business situation to a fight with our spouses. It's an honest fact that we aren't the same every day: athletes aren't, and they play, fight, or compete better on some days than others. It is difficult to stand off and objectively analyze how we feel. But some days you've got it and some you don't and the ones you don't should, under some conditions, raise your ceiling and visibility limitations.

I can remember an all-night flight from Kansas City to New York with stops at St. Louis, Indianapolis, Dayton, Columbus, Pittsburgh, and then Newark. It was back in DC3 days, and the weather was bad with an instrument approach at each station down through an icy overcast to minimums. The flight was a heavy mail flight, and at each station we were delayed while they loaded and unloaded mail. I had a new copilot, too. Finally, 5 A.M. found me in Pittsburgh looking at Newark's weather, which was forecast poor with precipitation. I turned to the agent and told him the flight would hold for six hours and quickly added that it wasn't that the weather was too bad, but only that I was exhausted and didn't think it wise to fly any more bad weather without some rest. The airline, incidentally, never complained about it, and when I dropped by the chief pilot's office a few days later to tell him what I'd done, he complimented me on the decision.

✝ More about Wind

We've looked at the forecast, the ceiling, visibility, precipitation, and at how we feel...what else? Wind direction and velocity in relation to terrain, such as light winds flowing from bodies of water or strong winds flowing down nearby mountains. Some airports, snuggled beautifully against the side of a mountain, can be a tough place to land with high surface winds. Hidden in the winds can be the rotor from wave action downwind of the mountain. It may be only a few hundred feet above the ground and have very nasty turbulence plus areas of strong sink that will have you pouring on power while wrestling with the rough air. The trick is to be suspicious of such things close to mountains and not be caught off guard.

There are details, too, of runway alignment. Now and then we may be headed for a field with a single runway where a crosswind will be too much to handle.

High-wind velocity is an indication of turbulence in a low approach, which will make the approach more difficult.

Calm winds and approaches are delightful; they are easy approaches even though, quite probably, they will be low ones with reduced visibility because of the calm air.

Speaking of airports, we should check NOTAMs for the destination. Has the snow been removed? Are the lights and facilities all working? What other things are affecting the airport's condition?

✝ Altimeter Setting

Altimeter setting is reported on the weather sequences, and as we look at present and past reports, it's a good thought to make mental note of the pressure and see if it's increasing, decreasing, or staying the same. We can catch a weather trend this way and, keeping in mind the pressure, we have a reference for altimeter setting in case we don't hear one along the way. This is not recommended for IFR flying, for which an accurate altimeter setting is needed all the time; but the information is a little backup "just in case."

✈ Temperature and Dewpoint Again

On those weather sequences, we find our old friends temperature and dewpoint. We want to keep them in mind to relate to the present and forecast weather, and as a reference when watching the trend at our destination. It's important to note, again, that time of day affects the value of temperature dewpoint: sun coming up, they tend to separate; sun going down, they get closer with chances for fog; they come together when it starts to rain, too.

✈ Pilot Reports

Pilot reports are worth careful consideration, because another pilot reporting the height of the tops, for example, is the next best thing to being there oneself. Like almost everything else, however, these reports must be absorbed with some thought about their validity. A report of the tops is a pretty definite thing, but reports of turbulence can be questionable. How turbulent is it, really? One pilot, being the nervous type, may yell severe turbulence while old hard-nosed Joe says it's just choppy. Notice, also, what kind of an airplane the report came from—one with a high wing loading wouldn't be as troubled as a lighter airplane of the two- or four-place variety. So, what do we do if turbulence is reported? Look over the general situation and decide if there should be turbulence; if there should and it's reported, that's that. If there shouldn't be turbulence and it's reported nevertheless, then either a Nervous Nelly was flying the reporting airplane, or something has started to move in the weather pattern that's different from what's forecast, and we'd better look things over again and decide what it might be. In either case the turbulence report didn't hurt, and has only caused us to study things a little more and be prepared.

As pilots, we should make an effort to report the weather we find, particularly cloud tops and bases, icing, turbulence, thunderstorms, or anything unusual. These reports not only help pilots directly, but they help the meteorologist, which finally helps pilots and a lot of other people, too. Meteorologists have information on places that report weather, but they do not know what's in between. Sometimes weather will cook up something they never knew about. A pilot has a chance to see these in-between places and can realize that something different from the report may be happening. This is the time—an important time—to make a pilot report and to let the meteorologist know, too. A report that says, "Snow encountered between Albuquerque and Gallup," may be an important clue that things are happening faster than expected, and the meteorologist had better revise the forecast.

☦ On the Ground, Too

Report the braking action you find on landing when ice, snow, water, or whatever make stopping an adventure. Pilot braking action reports are often the only information available for an airport—so your braking action report is important.

Remember the reverse, too, and request past pilot braking reports for your own information before landing on questionable winter days and nights—or times when the runway is water covered, perhaps the result of a thunderstorm or prolonged heavy rain.

☦ Summing up

For the terminal, then, we study forecasts for past, present, and future; we read old and current sequences and check them against the forecasts to see how the forecasters have been performing. We look at other terminals within a couple of hundred miles of our destination and relate them to our destination weather, and most important, we study reports and forecasts of airports beyond our destination from which weather might come. If we are westbound and a front is approaching the field for which we're headed, we check how the front has been acting as it approaches. Say we're going from Columbus, Ohio, to St. Louis; in addition to St. Louis, it's important that we study Kansas City; Springfield, Missouri; Oklahoma City; Burlington, Iowa, and a station or two west of there, like Omaha. A wide sweeping glance will tell us where the action is, and then we relate it to St. Louis.

We need an alternate for our destination airport, even when flying VFR and especially marginal VFR. We study alternates as thoroughly as we do the destination, and perhaps add a few factors. One is that we want the alternate to be as good and have as little chance of going bad as possible. While we may head for a destination that has a nervous forecast, we want an alternate that has a solid one, and we don't like alternates that haven't.

In tune with this we should always have, tucked back in our mind, which direction to go for an "out"—a diversion toward improving weather, the safest weather area. This is important for all flying, but especially VFR.

Alternate minimums are fairly low and it is possible to have an alternate that is forecast to be above alternate limits for arrival and below them an hour later. This brings about such statements from pilots as, "That's the paper alternate, but my real out is ___." They want something that's definitely going to be good, like an airport that is behind a front and on the uptrend or one that will not be influenced by the destination weather, like Montreal when the New York area has a low moving off the coast. Long Beach, California, may be a legitimate alternate for Los Angeles, but a knowing pilot says that Palmdale is the solid-gold out. It's wise to consider that the FARs give numbers to go by, but since weather isn't as precise as numbers, one needs extra protection on many occasions above and beyond the FARs. Paradoxically, or hypocritically, the FARs are often too restrictive and hold us back when it isn't necessary. However, this is the outcome of an interplay between the need for definite rules in the book, and the fact that so much of flying and weather depends on judgment. Judgment is difficult to cover in regulations. I'm glad I don't have to write them.

It's worth noting, before we leave terminal weather and the ways of looking it over, that we always relate our weather study back to the basics: big picture, time of day and season, temperature change, moisture change, terrain, wind, forecast, and the almighty "IF."

7

Checking Weather
for the Route

After considering the destination and deciding it's good enough, we have to learn what's in between: the en route weather. What bothers us en route? If we're VFR it's cut and dried: we must have enough ceiling and visibility to stay VFR, and the more mountainous the terrain, the more ceiling and visibility we need. We can fly on top of clouds, but they must be no more than scattered and very clearly and confidently forecast to stay that way!

On instruments our concerns are ice and thunderstorms, cloud tops, bases, and ceiling underneath, especially if we're single-engine and want some degree of tranquility regarding the possibility of engine failure. Turbulence is a factor, but is generally related to thunderstorms. Wondering about ice and thunderstorms, the important question is: Are we dealing with them in fronts or only air-mass conditions? This knowledge is very important because with air-mass thunderstorms your chances of steering around them are good; for ice you probably can get on top—unless the tops are too high over mountains. Through fronts, however, the thunderstorms will appear as a solid line that's impenetrable, and ice will be between you and the other side of the front with strong probabilities of confrontation.

VFR or IFR, of course, we want to know about winds—head or tail and how strong. But more of all this as we go along.

The big picture has its usual importance, and we want to take note of the weather system that may move onto or off the route. The method of looking over en route weather is much the same as for the terminals; we want the same scrutiny of forecasts for before, now, when we get there, and after. We mix in the same factors and emphasize terrain because it intensifies any bad weather that may lie along the flight path.

✛ Weather Is Mostly Good

I talk a lot about bad weather and may put a gloomy look on every-thing, but I'd like to make two points. One is that if we only talked about clear weather, there wouldn't be much to talk about. The other is that most of the time, thank heaven, the weather is good. We fly in clear skies more than we do in annoyingly cloudy ones. We may have cloud cover due to postfrontal conditions, but it is easy to top and fly above.

Fronts along the route, with their associated low-pressure areas, are the things that can make it tough. In these areas we find the diffi-cult weather, the ice or thunderstorms, thick cloud decks, precipitation, high-speed winds, and turbulence.

But this kind of weather covers our routes only a small percent-age of the time. I spent four and a half years in weather research, trying to find bad weather. I sat on the ground many, many more hours waiting for bad weather than I have ever sat as a pilot waiting for good weather. Often, too, the bad weather that was supposed to be out there wasn't, and instead of ice or snow I'd find myself flying disappointingly on top—looking at blue sky or stars. All this is simply to say that things are good more than they are bad, but we have to talk most about the bad.

Clouds rarely start at 200 feet and go up in a solid mass to 30,000 feet or so. We can draw a mental vertical section of how clouds are stacked. Usually they are in layers that become more complicated and the spaces between them fewer the closer we get to a low or front. In the simplest form we may have a single stratus deck behind a cold front, one layer with a reasonably reachable top or one layer where warm air is overrunning cold ahead of a warm front–a cirrostratus deck, for ex-ample, with a high base and nothing below. Then, closer to the warm front, we find an altostratus layer with a lower base. Closer in, this low-ers still further and rain starts, whereupon a low stratus deck forms, adding another layer. At the front, the layers merge.

✛ Something on Fronts

A cold front has much the same profile, but its area is smaller. The distance from front to back is shorter, and it does not take long to fly through the front. A cold front can, however, be more violent than a warm front. Warm fronts are slow—and sometimes tormenting—while a cold front is a quick punch in the nose.

As we know from primary meteorological study and from picture-book drawings, fronts come out of a low-pressure system. The warm front is a long arm sloping out ahead of the low, and the cold front is another arm pointed down, southward, from the low. The farther away from the low we are located on a front, the less violent the weather—and the closer to the low, the tougher.

☈ Occlusions and Zippers

A type of front we haven't talked about is the occluded front. I remember when I was a very youthful copilot who hung around the weather office on my days off, and I asked the meteorologist what an occlusion was. He said, simply, "It's just like closing a zipper." This really didn't tell me much until I learned more and found that his explanation was quite exact. At the center of a low the warm and cold fronts meet. When, near the center, they get together and one catches up with the other, they form one front. This process of catching up progresses downward from the low center. It's as though a zipper pull handle were at the low center, with one side of the open zipper the cold front and the other the warm front; the zipper closes—pulls—southward as the cold and warm fronts come together and become an occluded front. What does it all mean? Generally, it means the low is beginning to fill up and weaken, with cloud bases getting higher and tops lower, until the low is finally a trough with little weather. But in the early stages of an occlusion, things can be rough and tough, with the characteristics of a cold or warm front depending on whether the cold air is catching up to the warm front and lifting it upward as a cold front, or riding up the warmer air ahead to give it the qualities of a warm front.

Sometimes the "closed zipper" portion of the occluded front tips and bends over backward. It is then called a "bent-back occlusion," and will act like an additional cold front. Because of the counter-clockwise circulation about the low, such an occlusion is fed moisture and can cause some mighty mean weather. With the low near a coastline, a massive amount of moisture is fed back over the top of the low behind the occlusion. The low may move slowly, or not at all, and just sit there spinning around, moving that moisture back around it. Such a low, stalled over the Gulf of Maine off the New England coast, can bring massive snowfalls and foul weather for days.

Generally we think of an occlusion as "occluding out," which means that the low is starting to fill and die. Like everything in weather, however, you cannot always count on this, and while an occluding low generally spells gradual improvement, it can, on occasion, regenerate itself and create more bad weather. It's necessary to keep a close eye on occlusions. Visualize from where it may continue to have a moisture source to feed into it as a hint to what may happen. If it's in the dry part of the country, and a big moisture source is not available, then the low will fill and things will get better. But near an ocean or gulf, watch out!

✝ Large-area Weather

Some frontal conditions can be very widespread and cause weather over a large area. We find this most often in winter, in big lows having large areas of overrunning warm air in their northeast sector. In winter, these can cause large areas of freezing rain. Usually this type of storm is not difficult to spot or predict, and the Weather Service will have them sufficiently well analyzed to make an accurate forecast, but that doesn't relieve our task of keeping a skeptical eye on it. As far as flying into such an area is concerned, we shouldn't; it's better to consider the coziness of a fireplace and a good book, unless we're experienced, well equipped, and have lots of fuel.

✝ The Important NE Corner

As we study weather systems, it's important to visualize where the weather is. In a low, most of it is ahead of the system, in the northeast corner (in the northern hemisphere). This is ahead of and in the warm front. If we were to cross a system starting in the east-northeast and fly southwest, we'd first fly under high clouds and then into an area of rain or snow or possibly freezing rain, and then through an area of heavy precipitation, ice, and thunderstorms. This is the warm frontal surface. Then we would break out in a relatively clear area with a deck of cumulus or stratus or even clear skies. This is the warm sector. Incidentally, a low will generally move in the direction in which the isobars are oriented in the warm sector.

A weather system—the big picture. Note all the clouds and problems in the northeast sector. Thinking of this picture and then relating it to your location in an actual weather situation can tell you what weather to expect and which way to go. It's a picture that deserves a lot of study and mulling over. (Note the jetstream which would be up high, near the tropopause.) (NOAA PHOTO)

Past the warm sector we'd bump into the cold front: a narrow band of thunderstorms or ice, with lots of clouds and turbulence, but a fairly quick passage. Immediately behind a strong cold front there is a clear area, and we may think all is over, but as the cold air builds in, a stratocumulus deck will develop and in winter probably will become a solid overcast with fairly high tops and lots of ice in the clouds. But as we fly farther from the front, this cloud mass begins to show breaks, then becomes broken and gradually scattered with the tops lowering, and finally there will be clear skies. How fast all this occurs depends on our distance from the low center.

The interesting part of the system, however, is that northeast portion. In this area are the easterly surface winds that cause fog and low clouds. Toward the front, precipitation begins, causing low clouds and wide areas of poor visibility and ceiling. In winter, rain aloft will fall into cold air and create freezing rain and ice storms. This is also the area of heavy snowfall. This area may often be extensive enough to make finding an alternate airport difficult; it can also take a long

Drawing of a low taken from an actual weather map. Note the mass of clouds on the northeast and north side extending far west. Also note the east wind covering that entire area, bringing in the messy weather. Be wind conscious; where it comes from and what that source region is like and what weather it may breed.

time to move off. Although we cannot forget the cold front, the big action is up there ahead of the warm front, and this can be true hundreds of miles away from the low as well as near it.

In Dr. Horace R. Byer's book, *Synoptic and Aeronautical Meteorology*, he says, very accurately, "The warm front situation may properly be regarded as the most serious hazard to aviation..."

That's a good statement because a warm front has it all, and it's not simply the frontal surface itself, but also the big area northeast of it that requires study and respect when we look over the weather.

While studying the warm front and NE sector, it's a good idea to visualize being in various parts of this and which way we'd retreat toward better weather. The warm sector, south of the warm front, generally has the better weather, though we should watch that area for the eventual arrival of a cold front from the west. Running away toward the NE demands care to be certain the weather isn't running with us and preventing our escape from the lowering ceilings, visibilities, thunderstorms, and increasing precipitation.

Warm fronts slope, in a shallow way, from 100 to 300 to 1. So if the frontal surface—the place where surface reports show wind shifting from easterly to southerly—is at, say, Amarillo, Texas, a pilot will find the warm front aloft anywhere from 100 to 300 miles northeast of Amarillo. That would be the altitude where warm-front thunderstorms begin, or how high we'd have to climb for above-freezing temperatures when there's freezing rain down low and we want out of it. This warm-front slope is worth mulling over at leisure and important to note when checking weather.

✛ Go the Short Way

From studying weather, we tend to get the idea that we are always flying through fronts at right angles. Of course this isn't the case; sometimes our course is along a front, or at any angle to it, in which case tough weather can be prolonged. So there are times when it's wise to detour a bit and take on a front at right angles rather than slugging it out for a long period along the front.

The analysis of weather in lows and fronts tells what we may have to contend with en route; this is what we are looking for when we study the big picture.

Checking the en route weather, the VFR pilot decides if the ceilings and visibilities are high enough to fly safely under clouds. If the decision says they are, then the question is, Will they stay that way? The pilot wants to know if any system is moving in that will make the en route weather deteriorate. If there is approaching weather, but the pilot thinks there is time to complete the flight before it reduces the ceilings and visibilities below a workable limit, then the pilot must decide which way to run and what to do if the deterioration occurs ahead of schedule with visibilities going down and the ceiling forcing the airplane lower and lower.

We must remember, and it's very important, that the timing of forecasts is not always exact; they cannot hit precisely the moment when a front will pass, or a place fog in, or clear, or any other weather. The state of the art is not such that weather can be forecast all the time with absolute accuracy. That's why we must know weather and watch its movements all the time. I may repeat this theme many times, but I do because it is so important. We may be puzzled by weather's changes, but we should never be surprised or caught off balance!

The instrument pilot looks at the weather en route differently. She or he can accept weather that has ceilings and visibilities well below those necessary for safe contact flying. The question is, how tough will any turbulence, ice, or thunderstorms be in relation to the equipment and pilot ability?

✈ It Takes Time to Know

It would be easy to say that if pilots are going to fly instruments, they should know how to fly instruments extremely well and should have all the equipment; but this isn't very realistic. Even an FAA-rated instrument pilot has to go through a learning period. The rating, though earned after a lot of hard work, is only a beginning.

The FAA says, in effect, that rated pilots, new or old hand, may take off, go immediately on instruments, stay on instruments, fight ice and thunderstorms, shoot a 200-foot ceiling in heavy rain at the other end, miss the approach, pull up on instruments, and go to an alternate and make an approach there. That's a pretty exciting day's work for the newly rated pilot, especially flying without a copilot and busy getting ATC clearances and tuning radios.

It's obvious that the less experienced pilot doesn't want to fight all that (sometimes the experienced one doesn't either); so what does the new pilot do? Learn to crawl, then walk, and finally run; that is, take on weather a little bit at a time, gaining experience with each flight. (More about this later.)

Seeing a weather map is an important part of studying en route weather, but often a map isn't available. Then we must get all our information from forecasts.

These start with a synopsis, which demonstrates that the Weather Service and FSS like to begin as we do—by looking at the big picture. Then they go on to tell cloud bases and tops, where icing and turbulence will be, and what the outlook is for the next weather period.

We read the forecast carefully and try to visualize where all this is going on. We try to make a dull bunch of words and symbols come alive and paint a picture. It may say "—22020G30KTS 3/4SM SN VV012," which doesn't seem to make much of a picture; but it can if we concentrate.

The ceiling is going to be indefinite 1,200 feet, which isn't bad; but three-quarters of a mile visibility gives a gloomy sight of flying in severely reduced visibility. We'll see only straight down and very little ahead, especially in snow. The ground will be snow-covered and all look much the same; navigation will be difficult. If the terrain is hilly with mountains nearby, we'll have to be extra careful. It's not a VFR operation.

The report continues to say that the ceiling occasionally will be 300 feet, one-half-mile visibility in snow showers. We picture sudden deterioration in conditions as the wind picks up and heavier snow showers come in gusts. Your skeptical nature tells you that the visibility will not be one-half from the cockpit, but more like one-quarter mile. We know it will require an instrument approach to the runway.

Probably one of the most important points in these forecasts (TAF) is the probability outlook stated, for example, as PROB40 0407—which means there's a 40 percent probability between 04 hours and 07 hours that something will be better or worse, the "something" being indicated by the coded letters. The TAF might also say, "BECMG 0407 __," (Becoming between 04:00 and 07:00.) What it will become will be expressed in the code; FG for FOG, SKC for SKY CLEAR, and so forth. What the coded expression PROB40 tells you is that there is some question as to the preciseness of the forecast, that there's some doubt, which is a signal for you to be extra wary, and to watch and keep up with the weather's action.

The statement BECMG 0407 means that between four o'clock and seven o'clock things will change for better or worse. It's your job to see how that prognostication is performing; if it's supposed to become better in that time period, it's your task to watch and see that it does—which means the old business of getting the en route weather reports each hour and writing them down in order to compare the latest with earlier ones and learn which way things are going. Don't be bashful, either, about asking by radio for the latest forecast or any supplemental ones that may have been issued.

Important in our en route flight are the winds aloft. First, of course, we study them to find their direction and velocity for flight-planning purposes, so that we know how long the flight will take and how much fuel we'll need. Add 20 percent to headwind velocities and take away 20 percent from tailwinds as a "fudge factor." That's about as accurately as they are forecast on an average.

The winds should be studied at various altitudes. We are generally accustomed to the idea that the winds are stronger the higher we go, but this is not always the case; sometimes we find lower velocities up higher—which could help if we were flying into a headwind. We study the wind at various levels in order to choose the best altitude at which to fly. An old rule we used in DC3 days when flying into head-winds was to fly high in southwest winds and low in northwest to reduce the wind effect. It's not always true, but it's not far off.

Wind is really a secondary factor in selecting altitudes; weather is first. We wouldn't want to fly in a deck of icing clouds merely in order to have a better tailwind or less headwind. So weather is really the first factor in picking an altitude (besides terrain clearance, of course, which comes before everything), and wind is second. Air-plane performance is another factor. We want to cruise as close to the airplane's optimum cruise altitude as possible; weather and wind will dictate whether we can or not.

It's good to keep in mind the wind at other levels, in case ATC sends you to a different altitude than the one you want. The trip might take longer at a different level. How will this affect fuel reserves? In a jet this could be quite serious, but it's important in a 100-horsepower light plane, also.

The winds aloft also give weather clues. Suppose the wind for the trip was forecast southwest, but as you fly you find the wind really coming from the southeast. That's a clue that the weather may be starting to do something different from what was forecast, and it's time, again, to watch developments more carefully. Especially in the eastern half of the United States pay special attention, if the wind works toward east, for generally deteriorating weather changes.

A last important point about studying en route weather regards flying around weather. We should not look only at the weather on the direct line between where we are and where we want to go. The old idea that the shortest distance between two points is a straight line doesn't always apply in flying. We should look well to either side of our route and see what's going on. Sometimes we can circumnavigate bad weather. Sometimes making an end run is worth it just to get better winds, because we are not interested in how far we fly in an airplane, but how long it takes. Going "off course" to get better winds may actually make the flight shorter.

This is especially true for long flights. If we're flying from the Midwest to California in winter, it's generally better to go via El Paso than via Denver; the mountains are lower, the weather less violent.

Even short flights sometimes can be flown in better weather by a small course deviation. It pays, in studying en route weather, not only to look on course, but on both sides, too. This applies particularly when there are thunderstorms rumbling about. The radar maps show the basic areas of thunderstorms and often we can go around the entire mess. However, I repeat that the radar maps are not for use to wiggle through a thunderstorm area, such action needs radar or spherics in the airplane.

✝ **Why and If**

Forecasts and synopses aren't enough; we need to know what the weather people are thinking deep down inside behind the official stuff that comes off the computer.

We should talk to a meteorologist if the weather setup is dicey. The place to do this is with a NWS station where real meteorologists live or, if you are subscribing to it, one of the special weather service companies that have meteorological staff. The reason we want to get this analysis is to know what the forecaster's confidence really is. What is the chance of the forecast working out as advertised? Is the forecaster 100 percent sure? Fifty percent? It's important for us to know. Getting a clue to their thinking not only tells us how cautious to be, but it also gives us a chance to ask "why" questions. If the ceiling is going lower, "why" is it? If the answer is because of an approaching front, then "why" is it approaching? That might result in us being told about a low-pressure system on the move, and then we ask "why" is it moving this way, how fast? The answer might be that a high-pressure area is blocking it from moving in another direction. And we could ask "why" is the high blocking and not moving? If we ask enough "whys," we can finally get a feel for how solid the facts are on which the forecast is based and thus how much confidence we want to put in it.

Asking "why" is a wonderful way to do this, and a great way to learn more about meteorology, too. Unfortunately, we usually don't have a meteorologist to talk with and have to rely on the published forecast percentages we spoke of before. If there isn't any way to ask the "why" questions, we can ask them of ourselves and try to figure out the answers as a good way of evaluating what may or may not occur.

After "why" comes "if." As we absorb the forecasts, the big picture should form a background in our mind into which we fit the forecast, and while doing this we develop our "if" thoughts.

"If" thinking is very important as we say to ourselves what "if" the front slows down, what "if" it speeds up, what "if" the wind stays east, what will it be like and what's the effect on me "if" the forecast does a 180-degree flip? And we ask ourselves, "if" it does something unexpected, what will I do in flight? Being "if" conscious keeps surprises out of life and has us prepared for something worse that might come along.

A simple example is a cold front in eastern Ohio. We might be flying from Bridgeport, Connecticut, to Harrisburg, Pennsylvania. The forecast says Harrisburg will have scattered clouds, visibility five miles in haze, and southwest wind. It will stay that way until 17:00 when the scattered will become broken with a risk of thunderstorms. The forecast for later shows thunderstorms and a frontal passage.

It's no problem for us because we are going to arrive there by 14:00, long before the thunderstorms. But our "if-mind" says, "Keep an eye on it." The front is out there and moving; a prefrontal line squall could pop out ahead of the front. It will be well to get the latest weather before leaving Bridgeport and current weather as we fly en route.

All this example does is demonstrate the kind of thinking a weather-wise pilot does, especially in relation to that synoptic picture.

In this simple case it's obvious the forecaster based the terminal estimate on the movement of the cold front. It looks as though it will move, but will it move at the rate expected? Best information says it will, but as we know, they cannot hit frontal movement on the nose. The forecaster planned for that a little in the report when it warned of a risk of thunderstorms for 17:00 in case things moved faster than expected.

If we were coming from Dayton, Ohio, to Pittsburgh, in this case we might be expecting Pittsburgh to be in good weather behind the front, but if it slowed down, Pittsburgh would not clear out and the front could still be in the area for our arrival. It's all part of the necessary suspicion, the constant knowledge that weather forecasts do go sour.

So when we look at the weather we keep in mind several key points: big picture, time of day and season, temperature-change possibilities, moisture-change possibilities, terrain, wind, forecasts, and the big WHAT IF!

✝ Don't Fear Weather

All the emphasis we've put on weather so far may make it sound as though the sky, most of the time, is full of fearsome conditions. Well this isn't so, and even when bad weather prevails, fear shouldn't be the attitude—rather it should be respect.

Weather makes flying interesting, even though it frustrates us at times. But if we fear it, our flying will be affected; we will not perform as well as we can and emotions may overcome judgment. It is fun, more often than not, to fly in bad weather. It makes us feel a part of nature again and not a coddled creature living only in soft comforts; it builds our confidence. The point is that to handle weather we must know, again and again, that it is capricious and its movements unreliable. What we are trying to do is make certain that we outguess this uncertainty and, when we cannot do that, be prepared to handle it. If we prepare by obtaining weather information before flight, keep up with it in flight, and have alternate action always available, then we have respected weather and can handle it. But if we do not do these things, we may well find ourselves in an alarming position, scared of everything around us and not thinking clearly—a recipe for disaster!

✝ Or Worry About It

Worrying about weather shouldn't upset our lives, and it doesn't help to fuss in advance over what it's going to be like when we fly. In my youth I fretted about weather days ahead of a cross-country flight: Would there be ice? Thunderstorms? Low ceilings? Using a little extra imagination, it was easy to get all upset.

But somewhere along the way, I realized that there wasn't any way I could confront that weather until the day came to fly it—so there wasn't any point in stewing about it in advance.

It's good practice, of course, to check weather daily, in moderate fashion, to keep the trend in hand, but the time to get steely-eyed serious is when first walking in the briefing office or picking up the telephone to obtain the information. Then, with a good briefing and thoughtful preparation as to fuel, alternates, and how to approach what is out there, we can deal with the conditions in a relaxed and unworried mood. But worry the night before? Never! That's the time to get some sleep and be ready for what comes on the morrow.

8

Equipment Needs for Weather Flying

It is almost impossible to talk about flying weather without talking about the equipment of both the airplane and the pilot. The airplane we fly must fulfill certain requirements, not only of instruments and radio, but also of power and range to battle the elements when necessary. This doesn't mean it has to be a big four-engined transport, or even a small two-engined one; but the airplane should have a decent rate of climb, and more important, it must have enough range for the flight, from takeoff to destination, to be made against a strong headwind and along the wiggling course that may be necessary because of ATC. There is an impressive difference between the straight-line distance between two places and the distance measured over the wandering routes of the Federal Airways System.

✛ It's Farther than You Think

While the shortest distance between two points is a straight line, airplanes, flying instruments over the airways, don't fly that way. One flight I have often made is a 251-mile direct flight from Van Sants, a small field in Pennsylvania, to Montpelier, Vermont. If I filed instruments, however, the airways distance is 304 miles; that's over 50 miles extra, and in a Skyhawk that's a lot.

A flight made around the New York area was 60 percent farther by ATC-cleared route than over a more direct filed route that also avoided the New York airports area. The big extra distance takes lots

of additional fuel, and that affects safety. Suppose we were trying to beat darkness and approaching weather—a slightly stupid
combination—and then found our ATC clearance via a route that
would take 70 minutes longer, not counting possible headwinds. We
could easily run low on fuel, fly into darkness, and meet bad
weather head on! If there's any doubt, check the preferred route
with ATC in advance via FSS or telephone to the center.

Range also must include enough fuel for reasonable holding at
the destination and then diversion to an alternate airport against a
headwind. Furthermore, of course, some fuel should remain in the
tanks when arriving over the alternate.

This is the needed range:

1. Departure to destination, plus
2. Holding fuel, plus
3. Fuel to alternate, plus
4. Reserve over the alternate

Let's break these down further.

1. Departure to Destination   Departure to destination, the long
way. It is most important to know the en route time, with winds taken
into consideration. If we are going through bad weather, the wind
may be strong. Slugging against the winds of an active low-pressure
area may give ground speeds low enough to cause apprehension.

We need fuel for departure and arrival wanderings as directed by
radar. You do not simply fly to the destination's radio aid, make an
instrument approach, and land. If it's a busy terminal you may be
told to fly vectors that add many miles to the distance already flown
over the wandering airways. One route into JFK from the north has
an average vector distance of 91 miles! If we're flying a jet, and doing it at a low altitude which we may have been cleared to, those
91 miles can use up a lot of fuel and make our reserve a considerably, nervous, lower amount than we'd like to have. (Sometimes
these vectors carry one over big areas of water in the vicinity of
coastal or lakeside cities; this makes buoyant cushions that can be
used as life preservers a good piece of airplane equipment. Life vests
as regular equipment are a good idea; they don't take up much room
and weigh little.)

Fuel consumption is higher and speed lower during climb, so
the altitude to be flown will affect the total fuel needed. This can be
determined with a knowledge of the average rate of climb, the average true airspeed during climb, and the fuel consumption at climb

power. As a ready reference in flight planning, a pilot can make a handy chart that will give the fuel burned and the miles made good by the airplane in climbing to various altitudes. I have one stuck up on the side, slightly above head level, in my airplane—along with the checklist.

✠   Weather Uses Fuel

In icing conditions the airplane can lose speed, and because of extra power and/or carburetor heat, the fuel consumption will go up. Wandering around thunderstorms also adds miles and consumes additional fuel.

It is difficult to say with precision how much to add for ice or detours around thunderstorms, and pilot judgment must prevail. But awareness of the fact that these things take fuel is a good beginning, and a 10 percent cushion will serve as a start.

2. Holding Fuel   Again, it is difficult to judge how much we need, but there are factors to help us make an educated guess. First, Is our destination a busy terminal? If so, the delays can be long. Obviously, if we are going to Chicago's O'Hare, the delay will be longer than for Reading, Pennsylvania.

Time of arrival may also affect holding time. Any big terminal will have much longer delays between 4 and 8 P.M. than at 1 P.M. or midnight.

Holding ability depends also on the airplane. A piston-engine airplane can get its fuel consumption down quite low when just lolling about. A jet, however, uses lots of fuel and is especially critical at low levels. If jets can hold above 20,000 feet, their fuel consumption can be brought within reasonable limits. Fanjets are less greedy than straight jets, however, especially at low levels.

The weather forecast enters into the holding problem. A person doesn't want to hold, for example, while waiting for the arrival of a cold front. If ATC says the hold will be for an hour and a cold front is due within that hour, and if reports indicate that the forecast is accurate, it may be wise to proceed immediately to the alternate and forget about holding. If it is late and fog has formed, there is no point whatever in sticking around. Which, again, emphasizes the importance of keeping up with a range of weather reports in flight.

As a yardstick for the *minimum* holding fuel, there should be enough for one and a half hours at a busy terminal and 30 minutes at more out-of-the-way places. Traffic can back up badly at a busy terminal or alternate because of thunderstorms, during which landings may cease entirely. Traffic keeps pouring into the area, and stacks up even worse than in winter conditions. I've seen Kennedy go from 15-minute delays to two-and-a-half-hour delays in a matter of minutes because a thunderstorm hit the field: *it* had number one priority!

3. Fuel to Alternate   Fuel to alternate is just like fuel to destination and should include the same factors—i.e., total "long" distance, wind, and weather en route. Fuel consumption can be computed on a long-range-cruise basis to save fuel. If there is a very strong headwind, however, long range may not save fuel. But it's surprising how long-range cruise will save fuel even into a fairly strong headwind. A little study using computer and airplane manual—at a leisure time—will show this.

Headwinds affect airplanes according to their true air speeds. A 50-kt headwind against the Concorde is about 3 percent, or something like six minutes' extra time on a Paris-New York flight of 3,200 nautical miles. But 50 kts on the nose of a Cessna 172 is 45 percent, which means an hour and nine minutes extra between New York and Washington, about 175 nm!

It's important to explore headwind and tailwind numbers and to know what they do to range and fuel on our individual airplanes at normal and long-range airspeeds and different altitudes. With this knowledge, we can quickly set up for maximum range when we really need it.

4. Reserve over the Alternate   Reserve over the alternate depends on weather at the alternate. If it's clear and forecast to stay that way, the reserve can be minimal. If the alternate isn't clear and instrument approaches are required, even with a fairly high ceiling of 1,500 feet or so, more fuel is needed because there may be traffic delays there too. Often, when an airport closes down or becomes unusable, everyone holding at that airport flies off to the same alternate. This moves the congestion from one airport to another, and the alternate can be a madhouse of delay and confusion. Arriving over the alternate with minimum fuel in this situation is a scary proposition.

As we begin weather flying, of course, we should pick alternates that are clear and definitely forecast to stay that way. When we say "clear," we mean broken clouds or better and ceiling 2,000 feet above the highest terrain within 50 miles.

✝ Fuel and the Law

Alternate fuel reserve should be 45 minutes—the Federal Aviation Regulations say at normal cruise power setting—but it's good to keep in mind that this is the final fuel, and when the last of it slides through the engine, all problems become simple. You are going to land, right now, wherever you are!

The government regulations spell out fuel reserves, but these certainly should be considered minimums. This is all they ever could be because it's impossible to write regulations that cover all conditions— especially considering the fickleness of weather. Having sufficient fuel is one of the greatest safety factors one can give oneself. It assures the mental tranquility that is of paramount importance in weather flying. A pilot running low on fuel may cause a hurried emotional decision that is wrong.

It's worth reflecting that fuel management is a major cause of engine failure. Good management means, for one thing, having enough fuel—not running out! It also means having the good sense to be certain the fuel valve is always on a tank with fuel in it. Airplanes have run out of fuel and made forced landings with the fuel valve on an empty tank, while fuel remained in another tank!

The added advantage of being able to accept a wandering route from ATC, without being unnecessarily concerned about fuel, makes weather flying an interesting, enjoyable experience rather than a nervous and jittery ordeal. The ability to fly out of the weather area and go where it's clear is a solid comfort of the first order. A pilot should never be in a position where all the bridges are burned and there is no way out. A fat fuel reserve goes a long way to provide the necessary out. Desperation due to a dwindling fuel supply has undoubtedly caused more weather accidents, by far, than engine failures.

✝ Fuel Again

There are two ways to deal with fuel. One is to have lots of it—and airplanes are getting better every day in this regard. Certainly if maximum utility is desired, an airplane's manufacturer should strive to provide for the greatest possible fuel capacity when the airplane is originally designed.

The second is to fly within the airplane's fuel capacity by limiting the length of flights. This is restrictive, of course, because sometimes weather covers a big area that doesn't allow a short-range airplane any alternates. Then there's only one thing to do: sit and wait for better weather.

✛ Instruments

What else does the airplane need? Instruments to fly by, arranged in a good useful fashion. The hit-or-miss manner in which instruments are often stuck about cockpits is appalling. Good instrument flying requires constant visual scanning by the pilot. Logically, then, the shorter the distance a pair of eyes needs to travel to scan an instrument panel, the better.

Each panel arrangement presents its own problems. The space available, control-wheel location, windshield height, and other such factors dictate how instruments can be arranged, but regardless of these problems the flight information instruments need to be grouped closely together.

The basic key of instrument flying is to keep one's eyes constantly roving over the important flight instruments. When a pilot flies weather, the technique of flying by instruments should be so well developed that almost all attention can be devoted to the problems of weather, air traffic, and flight management. Ideally, an automatic pilot does the job, but not all airplanes are equipped with them, and generally during low approach and takeoff the pilot is not using the autopilot—except the modern airline aircraft in which the autopilot is used for landing a big percentage of the time.

Autopilots have reached a level of sophistication that benefits all aircraft tremendously. With modern big aircraft, autopilots can make approaches, kick out crab, land, and roll the aircraft to a stop, as well as do all the flight chores such as climb, descend, and follow navigation signals, be they simple omni, GPS, or Inertial. Smaller single- and twin-engine piston aircraft have autopilots that follow headings, track omni, fly ILS, and hold attitude, plus other good things. These are light and reliable because of today's electronic wizardry.

The autopilot can do the manual labor, but it can fail, too—not simply a total, doesn't-work failure, but a subtle failure like tracking a course improperly; and the chance the pilot didn't program it properly, or it sneaks off altitude hold, and many more combinations. All this means a pilot should frequently check that the autopilot is behaving properly. This is a matter of scanning against raw data shown on attitude, altimeter, IAS, and omni-ILS instruments. Because the autopilot relieves the high concentration of manual instrument flying, scanning becomes easier and better.

In the complex world of instrument flight, an autopilot is almost a requirement; someday they will probably be standard on all except pure sport airplanes; they are certainly very desirable, as they turn a busy hand-flying task into a relaxed experience where the pilot is well in command of the situation and always ahead of the airplane. The lower a pilot's experience level, the more an autopilot becomes a serious need, or, of course, a good copilot.

The key instruments are the artificial horizon, for bank and pitch reference, and the directional gyro for heading. The directional gyro is fast being replaced by the horizontal situation indicator (HSI), which is a big advance and help because it has heading and navigational information all together; the heading is slaved to a remote pickup and stabilized so one doesn't have to constantly worry about resetting the DG. Airspeed, vertical speed, and altimeter are really reference instruments, and their action is a result of things that appear first on the artificial horizon and directional gyro.

If the key instruments are watched closely, the airplane never gets a chance to go very far from the "straight and narrow." Frequent scanning and immediate correction of excursions from the desired flight path give the feeling that one has the airplane in a narrow corridor, boxed in on all sides, unable to escape and fly off on its own.

Since rapid scanning prevents the airplane from getting far off course, any correction the pilot needs to make will be small and therefore easy. Problems start when a big bank or pitch angle is allowed to develop. If this happens, the heading, vertical speed, airspeed, and altitude go off, and the pilot has a handful trying to get the airplane back where it belongs. Scanning is not difficult if practiced. Small corrections and never allowing the airplane to wander far make flying simple and relaxed.

An overlooked instrument is the vertical speed indicator. If it sits on zero, there cannot be much wrong with the airplane's flight; vertical speed cross-checked with the Directional Gyro is an easy, quick look that things are okay. A good vertical speed instrument with its indicating needle of the "old-fashioned" size—commonly referred to as a "steam gauge" type, is a valuable instrument to catch your eye when it starts up or down, if it does it's time to get on the other instruments and see what's going on. The vertical speed indicator is an underrated instrument. My awareness of it started years ago, probably as a result of glider flying where vertical speed is a most important instrument. My scan glance includes it almost as the first thing to look at—along with heading.

Small simulators, such as the desktop types available, are excellent trainers to improve one's scan and keep it up to speed. They improve our ability to handle complex ATC procedures and are an answer to the question of how to stay proficient when the autopilot is doing all the work. I have one and use it during those bad days of winter when I don't go flying very often. It keeps me current, and it's fun.

It is very important that the pilot be well versed in attitude flying. There are reams of information and instruction available, and an instrument pilot should take advantage of them to become a good attitude pilot. Attitude flying, in one simple example, is the difference between attempting to maintain a specific airspeed by chasing the airspeed indicator, rather than by keeping the horizon bar in a position that will give the specific airspeed desired. If speed is too high, the nose is raised slightly with reference to the horizon, thus producing a small but positive airspeed correction. Practice soon teaches what attitudes will provide the desired airspeeds, heading changes, descents, and climbs in small, easily controlled increments. But more important, it will help keep airplane movements small and prevent large oscillations that are difficult to recover from. Well-arranged instruments, combined with good scanning and attitude flying, make instrument flight precise and simple.

✝ Equipment Needs

Among the other things airplanes should have when flying instruments is pitot heat. Airlines feel that pitot heat is so important that they use it whenever they are flying, cloud or no cloud, summer or winter, to be certain it's always on in case of need. A temperature drop can take place inside a pitot tube and produce ice or slush if there's visible moisture. Even water from heavy rain can cause an erroneous reading in pressure instruments. Ice, of course, can knock out an airspeed indicator completely. Pitot heat is a must!

✝ We Can Keep it Simple

What kind of instruments does one need? The extent and sophistication of the instruments is determined by the amount and kind of weather a person is going to fly. A pilot can fly some pretty awful weather with quite simple instruments, but that's a very busy pilot, and the job accomplished won't be a neat, precise one.

My own instrument flying began with an airspeed, vertical speed, and a turn and bank. I taught myself how to use them by following a book called *The 1-2-3 System of Blind Flying*, which was written by an airmail pilot named Howard Stark, one of the important pioneers of instrument flying.

I flew actual instruments, too, but it was all restricted to climbing up through or flying on instruments toward clearing weather. It was done without radio by simply holding a heading. This wouldn't be possible today in most parts of the world because of airways and traffic, but back then there wasn't any traffic, nor any airways either.

✛ A Little More to Do a Lot

From that simplicity I graduated to a Douglas DC2; the only addition to its instrument panel over my little Monocoupe's was an artificial horizon and a directional gyro. Along with this, there was, of course, a radio to follow beams and a loop turned by hand to read bearings which said the station was at one end of the bearing or the other, but not at which one, because the loop didn't have direction-sensing capability.

With this setup we flew weather, lots of weather. The landing minimums at Newark, New Jersey, in 1937, were a 300-foot ceiling and one-half-mile visibility! Takeoff minimums were 100 feet and one-quarter mile! It wasn't done very sophisticatedly either. We flew the radio beam toward the station, which was down in the Newark meadows near Elizabeth, New Jersey. We crossed over the center of the station, called, in the slang of the day, the "cone of confusion," at 800 feet, chopped the throttles, shoved the nose down, and descended quickly to 300 feet; then we looked into the black night for a row of red neon lights that led to the black runway. What's so surprising, in retrospect, is that we did it often and successfully. Of course, a DC2 could fly at 80 miles an hour; that makes a big difference from a jet that you don't get under 135 knots.

All this is to show that a lot of weather can be flown with a primary flight group, an artificial horizon, and a directional gyro. We did, in those times, do a lot of practicing using the primary flight group only; needle, ball, airspeed, vertical speed, and every instrument check we took included an approach using only the primary instruments. I'm afraid this isn't done as much as it should be today. Personally, I like the turn and bank indicator over the turn coordinator and believe the flight cues more true using the turn and bank, but perhaps that's all because I was "raised" on a turn and bank. When I bought my Cessna it had a turn coordinator, which I promptly replaced with a turn and bank. I just don't feel that the artificial horizon look of the turn coordinator is relating truthfully to turn.

Our old artificial horizons did about what today's can. The directional gyro was the bore, because like simple DGs of today, it didn't have any north-seeking ability and needed to be set to agree with the compass. Gyros precess and must be set frequently. They must be set when the airplane is level and not turning, and, of course, the compass must be settled down so that it's reading accurately. In rough air, while working on omni or ILS, this can be difficult to do. Some gyros precess more than others—depending, generally, on the condition of the gyro. If it is well maintained and periodically overhauled, and if the vacuum source, in the case of vacuum instruments, is set to give a constant value of vacuum, a gyro can hold a heading for 15 minutes or more without much attention. A point to remember is that once a gyro starts to precess, the rate of precession increases the more it precesses; so it's wise to reset a gyro before it gets too far off. It is wise, also, to rate your DG by keeping a record of what its precession rate is. Set it carefully on a smooth day, then fly for 15 minutes and reset it, noting how many degrees off it was from the previous setting. Keep a record, and when the rate becomes too high—about a 10-degree error in 15 minutes—it's time to have it overhauled.

┿ Things Can Be Better

In modern instrument flying, which requires precise following of airways, a DG may add appreciably to the cockpit workload. Instrument designers have cured the problem by wedding the directional gyro and the compass so that a north-seeking unit constantly keeps the gyro's heading adjusted for magnetic north. This is done in different ways by different manufacturers and given different trade names; but the important point is that once the gyro is set, the pilot need never worry about resetting it, and the heading information is always precise without the turning error, run ahead, and hold back of a magnetic compass.

This intelligent type of gyro has been one of the biggest boons to precise instrument flying, and it reached aviation technology at about the time that heavy traffic and more narrow airways called for better flying. This, now, is the HSI we talked about; it's one of the first "extra" items I would put on my instrument panel. (On the sophisticated, big airplanes, the heading comes via the IRS, Inertial Reference System.)

To sum up at this point: The minimum one needs for instrument flight is the primary flight group plus an artificial horizon and directional gyro. With this, one can fly instruments and shoot low approaches; with the addition of a slaved gyro or HSI, however, life becomes easier—and *much* easier with an autopilot.

✢ Even Better

Technology goes further. It has added intelligence to the other instruments. To explain a bit, if we fly an instrument landing system (ILS) with a horizon and DG, we fly headings toward the localizer; once on it, we try to find a heading that will keep us on it, by a process called *bracketing*. How much to turn and when to turn is a matter of skill and, to some extent, guesswork. As Dave Little, an American Airlines captain who has done a great amount of research in instrument flying, said, "Flying down a localizer this way is like following the white line on a road by watching it through a hole in the automobile's floor."

But now we have artificial horizons that have computed information fed into them. They discover how far off course one is, what drift there is, when an airplane is wandering from the course, and then say, "Turn now and this much." The information is presented in such a way that one has only to follow a little airplane, or bar symbols, superimposed on the horizon, matching one with the other, and by so doing a pilot stays right on localizer and glide path. It is called the Flight Director. The pilot still refers, although only periodically, to the basic localizer/glideslope instrument to make certain that the flight director is doing its job.

With equipment of this sophistication aboard, jets are allowed to shoot minimums as low as 200 feet. The flight director is required for ceilings of 200 feet and the autopilot below that. If the flight director is not working, minimums are higher; this is one reason why airliners carry duplicate systems. With this type of equipment, a 200-foot ceiling is easier to fly than a 400-foot one without it.

The computer data fed into these flight director systems are the same as that given to the automatic pilot when it is set up for making low approaches; if the airplane is equipped with both (as airliners are), one system is used as a check against the other.

With the Boeing 747 came autoland, and our minimums were reduced at certain airports to 1,200-foot visibility and 100-foot ceiling.

(Today they can go to zero!) The many approaches I made to these low minimums were easier, more relaxed operations than flying by hand to a much higher minimum. The autopilot does the work and you have lots of time to scan the cockpit and double-check instruments. You have an understanding of the entire action rather than being mesmerized by the few instruments that one uses to keep in the slot when flying by hand. When the runway lights come into view, there's no worry about sensory illusions and getting too low and below glideslope on the visual part of the approach because you're there! Right at the runway—and in a few seconds rolling down the center line on the ground. It's the way to do it. But always remember that during the modern, sophisticated approach, a pilot should double-check the raw data (what the ILS needles and altimeter are saying, plus scanning the flight instruments—called raw data) to be certain that the fancy electronic system is doing its thing properly.

✝ The Future Will Be Even Better

Now there are very sophisticated system-guiding airplanes such as information that tells us precisely and quickly what power is needed for each condition, and airspeed indicators that say at what rate we are increasing or losing airspeed. Computer systems, called Flight Management Systems (FMS), set up tracks, climbs, descents, and best power for cruise, a host of innovations in presentation as well as information; but the pilot, with a turn and bank, airspeed, and vertical speed indicator, will still be able to fly instruments. There will remain limits to the amount of instrument flying possible, just as the pilot with a little more equipment is limited, though not quite so much, and the one with everything is limited even less. The equipment, humble or extensive, must be used properly, with the pilot's proficiency honed to a fine edge through both knowledge and practice.

It is a good idea, too, for the pilot with the fancy equipment to maintain a certain proficiency with the simpler equipment. The basic flight group (airspeed, vertical speed, and turn and bank) are the most reliable, and they are always standbys in case other instruments fail. For more than 40 years, the turn and bank has been found on every instrument panel with which any instrument flying was done, from Cubs up to airliners and jet fighters. The turn and bank is beginning to disappear from airliners and some large aircraft, being replaced by a small standby horizon. But most general-aviation airplanes still have it; it's a good instrument and keeps working no matter what the aircraft's attitude is. If one is on your instrument panel, it's a good idea to cover the horizon and DG now and then and have a practice session with the turn and bank only. This is good simulator exercise, too. Remember that turn coordinators are not artificial horizons, but gimmicked-up turn indicators, and should be flown as such. They still retain the valuable property of working, regardless of attitude, however.

The computerized cockpit of FMS and flat-plate CRTs (cathode-ray tubes) are wonders to behold and do impressive things, but these systems are still subject to failure and, because of their complexity, the human can improperly program the system. Pilots, for a long time, will be required to monitor that all is well and the simple facts of where we are, how high in relation to terrain, and what the airplane's flight condition is. It's wise to remember that the pilot is ultimately responsible for avoiding the terrain and guiding the airplane on the path it is supposed to follow. Some computer people give the impression that airplanes can be flown entirely by electronic systems, doing a better job than the human. Quite true, but not in an unlimited sense. I often ask the question, "Would you fly on an airliner without pilots and only flown by computers?" The answer is obvious. So we must always be skeptical and double-check our situation.

☩ Power for Instruments

Instruments, however, are no better than their power source. Some are vacuum-powered and others electrical. There should be an alternate source for gyro instruments in case the basic source fails. This can be supplied in a number of ways. If the instruments are vacuum-powered, then there should be two vacuum pumps on twin-engined aircraft, or a standby artificial horizon or turn and bank run by batteries, such as we have in sailplanes. A turn and bank can be powered by an outside venturi. Although people shudder to think of venturis sticking out, causing drag and getting iced up. A system on my Pitcairn Mailwing, vintage 1930, took care of at least the latter problem well. The T&B venturi was mounted on an exhaust stack with five tiny holes in the stack just ahead of the venturi's entrance. These squirted hot exhaust gas over the venturi and prevented it from icing up. Such a rig might not be a bad alternate source today.

Most modern instruments are electrically powered. It is important to have for these an electrical system with two generators. Most batteries are just a wide place in an electrical wire and will not supply all the airplane's electrical needs very long on their own, although modern radios require far less power than old types. But even they will eventually use up the battery's charge.

There have been some dramatic situations and accidents because electrical systems lacked sufficient backup. If an airplane is going to be used for instrument flying, its electrical system should have a backup.

There are some wonderful solid-state radios that run from dry batteries and do so for a long time. My glider has a 720-channel two-way radio with good quality and power. I've run it over 12 hours on two small 7-volt batteries. One like that would be a big comfort in an airplane. It always would be there, independent of the electrical system.

The airspeed indicator and altimeter are "powered" by pitot and static tubes that bring in pressure and static air. Since it's very difficult to fly without airspeed, it's important to keep these sources clean and operating. We've already mentioned pitot heat. If the static source is within the pitot head, then the pitot heat will take care of it; if the static source is flush on the outside of the airplane, as most are today, it will not be closed by ice, but it can be blocked by slush thrown up from a nosewheel and then freezing, or by human-made things like tape put over the static holes during airplane washing or servicing. Careful inspection is necessary before flight to be certain that the static source is free and that the pitot tube is also. A bad static source will affect the altimeter, vertical speed, airspeed, and some pressurization systems.

The FAA requires that the condition of these instruments be checked periodically for leaks and deterioration of the tubes in the system. Instrument flying is important in weather flying, and while a person may or may not be proficient, all pilots can make certain the instruments they are using will be in top shape and functioning properly.

✛ Instrument Location

Almost as important as the condition of instruments is their location. The basic T system of layout is used worldwide. It has been proven over 50 years of flight and is found, in principle, even on the most modern transport aircraft instruments systems. This system positions the artificial horizon top center and the directional gyro or HSI directly below it. The altimeter is placed right of the horizon, and the airspeed indicator to its left. This location of the artificial horizon—the attitude instrument—in the most prominent and easy-to-use location is a testimony to the importance of attitude flying.

The important point is that the instruments should be as nearly in front of the pilot as possible, and they should be in clear view and not hidden by the control wheel or anything else. I flew the AirBus side stick system, and I'm enthusiastic about the location, as it leaves a wide-open view of the instrument panel; one could even have a laptop writing desk to keep notes.

✛ Lighted Well

It is important to have the instruments lighted properly. This means that at night the entire area of their dials should be illuminated without any shadows. There should be no glare, and they should be visible in a minimum of light. It should not be necessary to turn the lights up bright or to have excessively bright lights in order to see the instruments. Some airplanes I've seen require such bright panel lights to wipe out shadows on the instruments that outside vision is impaired. The pilot is like an actor trying to see the audience through bright footlights. The bright light not only cuts out what can be seen outside, but deteriorates a pilot's night vision. Bright lights also have a way of reflecting in windshields, so that as a pilot looks outside, the instrument dials grin back from the windshield or side windows in a most annoying way.

On the subject of night flying, it's worth realizing that night VFR flying is instrument flying to a large degree; while a noninstrument pilot can legally fly at night, a clear night with good visibility and some sort of horizon will be required. It can be an awful shock for a pilot to leave the end of a runway and to be suddenly enveloped by a pitch-black night with no horizon or reference to fly by!

If the horizon is not clearly defined and there is any risk of clouds, it's easy to slip into a cloud and never realize it until all outside reference is gone and we're on instruments! An instrument rating is a strong safeguard. The VFR accident rate is much higher during night flying than during day flying.

✛ Paperwork Is Equipment, Too

Almost as important as an airplane's instruments is the paperwork associated with instrument flying. We need maps, radio facility charts, and instrument approach plates, and they must be up-to-date. Note that I included maps, and I mean sectional type maps that show terrain: mountains, elevations, rivers, towns, airports, and all such. Too often the only chart used is the radio facility chart, which doesn't show the topographic features and a lot of other useful information. Reference using real maps is important for many reasons, such as avoiding mountains, locating airports in relation to towns, the water areas that may affect weather—a host of things. For safer flight, maps are needed.

But besides having up-to-date charts, plates, and maps, we have to have a place for them in the airplane. There's nothing worse than charts scattered all over the cockpit. Good instrument flying is good housekeeping. This means a little preparation in advance, before takeoff. The pilot ought to get out the charts needed for the flight, stack them in order, put them on a clipboard or some such device, and keep them close by so they are handy when needed.

Instrument plates should be treated the same way, and a clever pilot will have a clip or some such gadget on the control column where the plate can be displayed for reference while making an approach. It should be lighted.

The charts that we fly with are confusing and often difficult to read and interpret. There have been serious mistakes made because of the way in which things are sometimes presented. The papers we work from are a built-in hazard and should be treated as such. It is necessary to study routes and terminals in advance so you don't have to do it in the cockpit under poor light while bouncing around and trying to fly the airplane. (One way to catch errors that may result in using the wrong facility is to always identify what's been selected before using it!)

It's a good plan to have a systematic procedure for studying instrument plates. I always start with the field elevation, then the minimums, the ILS course, frequency and identification, the altitude of the glideslope at the outer marker, the minimum altitude I'll go to before pullout, then what the pullout procedure is in case of a missed approach. I look for the highest obstruction on the charts and note the minimum safe altitude within 50 miles. Then the runway length, lights, and approach lights. Also very important are the little notes that chart makers stick off in corners that say things like, "Do not descend below 4,400 feet on approach to runway 21 until Radar Controller advises passing 4,074 feet stack 5 NM NE of apt."

Now we have standard instrument departures (SIDs). They create another chart that must be consulted before takeoff. It is important not to be rushed into taking off until one understands what the SID says and has it firmly implanted in one's mind.

Similar to SIDs are STARs (standard terminal arrival routes), except, of course, the STAR routes are for arrivals and show the proper routing, altitudes, and sometimes speeds. Combined with STARs are profile descents for high-flying aircraft. Sometimes profile descents have their own special charts, too.

All these are important because there is lots of "fine print," which makes it easy to miss some ambiguously worded instruction. It's best to study these carefully in advance, on the ground and at leisure, and then to refresh one's memory before starting descent, preferably in a relaxed, cruise portion of the flight.

Rather than hunting all over the chart when busy in flight, felt marking pens to overlay routes, altitudes, or any other important bits of information are a big help. Yellow is a popular color, but almost any color allows one to see immediately the area being used.

✝ Go Fast Slowly

There's a great tendency at busy airports for clearances to be delivered too fast and followed by a rushed, "Take position; be ready for an immediate takeoff!" It comes in fire-hose fashion and tends to make people take off, half understanding how and where to go when in the air.

The answer is simple: to heck with the tower. Sit there and study the clearance. If you don't get it, ask for a repeat. Then, when certain, ask for takeoff clearance. It's better to delay the takeoff considerably rather than rush off half-informed.

In the cockpit we need a place to write clearances, then checkpoint times and estimates. We want to copy weather and clearance changes, keep track of fuel used, and write down a lot of other things. We need a clipboard or knee board strapped on, handy for the job. A computer (electronic or plastic) and plotter, big enough to see and handy at all times, are also indispensable pieces of equipment.

✝ Good Housekeeping

All this is part of instrument flying, and keeping a neat, organized cockpit makes instrument flying much easier. But more important is the fact that a disorganized cockpit makes instrument and weather flying much more difficult and increases the possibility of serious mistakes.

We like to have a copilot to help reduce this instrument flying load; but even if we don't have an official copilot (that is, a pilot with ratings), we can often train a nonflying person to look things up and sometimes even to handle the radio. This is best done by sessions at home when the pilot goes over the airways, approach, and navigation charts with the nonpilot and explains what they are, how certain ones are needed in flight, how to find them, how to add mileage, how to fold charts for pilot convenience and point to where they are, how to dig out frequencies, and many other small but useful duties. This can expand to learning to use a computer for speeds and fuel, and to limited radio work.

A little training of this type may turn spouse, child, or friend from a bored or nervous passenger into a useful, interested, and relaxed crew member, even if they don't know how to fly.

A copilot during serious instrument flight is a necessity—not required by regulation, but without a doubt, flying instruments without a copilot is a very tough job. I've flown small aircraft in serious weather, on instruments, alone and it's much more difficult than flying a 747 with a crew.

✛   Boom Mike

Without a helper, the workload gets pretty heavy. One way to reduce it is by using a boom mike. The idea of fumbling around to find the mike, getting it off the hook, holding it up and pushing the button, taking one hand out of use, is terribly primitive. With a boom mike and a button on the control wheel, one need only squeeze a finger to communicate. The other hand is free to write messages, adjust gadgets, and do the many chores it may be called upon to do. Boom mikes are thought important enough to be required by regulation on transport aircraft in the United Kingdom, and the FAA requires it on the single-pilot Cessna Citation. The wide use of noise-reducing earphones on many general-aviation aircraft has automatically brought the boom mike into wide use—a fine advance.

✛   Navigation

In addition to omni and radio beacons, we now find Loran C and GPS, Global Positioning System. These are excellent aids, but it's wise to remember that they have computers as an important part of their systems. When one has a computer, added effort is necessary to be certain, by care and double-checking, that it has been programmed correctly! Serious errors have occurred because of computer programming errors. This applies to inertial reference systems (IRS) as well.

Also remember that low-frequency aids such as Loran can be subject to precipitation static in heavy snow, some rain, and in thunderstorms. The best defense is to be certain the installation is properly done and the airplane has been rigged with antistatic antenna, discharge wicks, and such to minimize "P" static, as it's called. And realize that there may be occasional but infrequent periods of no reception because of it, regardless of the aircraft's equipment.

The new aids do such a remarkable job that we're apt to expect they do everything, but they do have limitations, and one should study the manuals, learn the equipment, and be certain to fly within the limitations.

The modern aids such as Loran and GPS have made navigation easy. They also have a tendency to lull us into a euphoric state of thinking everything is fine. Unfortunately, we hear too many cases of people crashing into mountains, not knowing where they were and unaware of possible flight path dangers. As I said before, it is still the pilot's responsibility to know where the airplane is, what terrain it's over, and how high that terrain is. Maps remain an important part of the cockpit gear; we have not reached the state where electronic gadgetry can be trusted 100 percent of the time. The picture of a pilot flying on autopilot, being guided by Loran or GPS and sitting back reading the newspaper, is a shocking one. Flying an aircraft should be a 100 percent job. Giving it any less attention is reckless.

☩ Radar and Spherics

For instrument flying and thunderstorms, we need the help of radar and/or one of the spherics units, but more on this later.

9

Temperature, an Important Part of Weather Flying

Temperature is closely related to many things we do with airplanes. It affects performance in takeoff, climb, cruise, and landing. It's important.

Air is the thing our airplane flies by. How much or how little air relates to the number of molecules. The amount of air striking the wing is a function of speed and of the number of air molecules present. If cold, the molecules are packed closely together, and we say the air is dense. If it's hot, there are fewer molecules per cubic foot, and the air is less dense. We sometimes say it's "thin."

The more dense the air, the better our peak performance. Engines put out more power; the wing has more lift in dense air. When the air gets hot and thin, it's all reversed and performance suffers.

✝ Temperature and Density

There's another way of looking at it. At sea level the air has a certain density, but at altitude the density is less. We know how this affects performance. When we take off and start climbing, our rate of climb is, say, 1,000 feet per minute. As we climb, the rate decreases until at the airplane's absolute ceiling, there isn't any rate of climb.

Why did the airplane stop climbing? It's simple: it ran out of air. The air wasn't dense enough to feed the engine and keep its power so it could push the wings fast enough to fly in the high and/or hot, less dense air. The airplane's performance is limited by thrust, or lack of it. If there isn't enough push, or pull, the airplane stops climbing. An unwary pilot may keep pulling the nose up, trying to eke out a higher altitude or stay at a marginal one and get to an angle of attack pretty near stall—and that's a fragile situation! So without air, an engine doesn't do its job, wings don't lift, and humans don't live.

This happens as we climb because the air is less and less dense at higher and higher altitudes—not because of heat or cold, but because the atmospheric pressure is lower. The higher we climb, the less air there is above us pushing down and creating pressure. It's always impressive to realize that half of the atmosphere is within the first 18,000 feet of altitude! At sea level, the atmospheric pressure is about 30 inches of mercury. At 18,000 feet it's about 15 inches of mercury.

Since hot air is less dense than cold air, hot air acts like air at high altitude, and cold air acts like air at lower altitude. What we've said is that air density and effective altitude are related. A sea-level airport on a hot day isn't at sea level at all as far as our airplane's performance is concerned. How high sea level can be sometimes comes as a surprise.

A sea-level airport on a day in summer, when the temperature is 38 degrees C (101 degrees F), has a density altitude of 3,000 feet! The sea-level performance of the airplane will have deteriorated about 20 percent. This figure will vary, especially with supercharged engines. In any case the airplane must not be overloaded; it must be flown carefully, and there must be enough runway for safe operation.

☩ Use the Computer

The density altitude can be found on any good pocket speed-distance computer. Most of them include a density-altitude computation window. It's worth a winter's evening to run out some make-believe situations and see how density altitudes can vary.

Airplane manuals generally show performance at different altitudes. The density altitudes we find on the computer can be compared with the airplane's performance at that altitude to see what performance we can safely expect on hot summer days. In the computations we must consider load—how heavy we are—and the advisability of reducing load by removing that camping equipment, "stuff" purchased, baggage, even a passenger or two. The airplane manual will show the performance versus load, temperature, altitude, and the lot. It's important to study and refer to it carefully. Grasping the control wheel with an upward pull, sitting forward on the edge of the seat, dry of mouth

U.S. Standard Atmosphere

Altitude	Temperature F	C	Pressure in. Hg	Millibar level (MB)
0	59	15	29:92	
1,000	55.4	13	28:86	
2,000	51.9	11	27:82	
3,000	48.3	9.1	26:82	
4,000	44.7	7.1	25:84	
5,000	41.2	5.1	24:90	
6,000	37.6	3.1	23:98	
7,000	34	1.1	23:09	
8,000	30.5	−0.8	22:23	
9,000	26.9	−2.8	21:39	
10,000	23.3	−4.8	20:58	700
11,000	19.8	−6.8	19:79	
12,000	16.2	−8.8	19:03	
13,000	12.6	−10.7	18:29	
14,000	9.1	−12.7	17:58	
15,000	5.5	−14.7	16:89	
18,000	−5.2	−20.7	14:94	500
30,000	−48.0	−44.4	8:88	300
34,000	−62.3	−52.4	7:38	250
36,089	−69.7	−56.5	6:68	Tropopause
39,000	−69.7	−56.5	5:81	200 (actually 38,600)
44,600	−69.7	−56.5	4:40	150

Note above tropopause: In the stratosphere the temperature remains constant at −56.5° C. This is standard, of course, and can vary because the tropopause height will vary widely.

and wishing the thing would climb and get over those trees is "too-late" time; the manual reference should have come some time before.

✛ How Hot, How High?

If one flies out of high-altitude airports, 5,000 feet or more, with hot temperatures, it's not unusual to have a density altitude of over

8,000 feet! Some low-powered airplanes will hardly fly at that altitude, and even the best will have impressively reduced performance.

Remember, standard temperature at sea level is 15 degrees C (59 degrees F). That's what all those manual figures are based on. The moment the temperature has gone one degree above 15 degrees C, the performance has started to deteriorate. Remember that on a warm summer day when taking off from a small country airport with trees to clear at the end of the runway!

Only 10 degrees above standard can make an impressive difference, and if the wind is light, or almost calm, don't shrug it off as unimportant and take off on any convenient runway. Be sure, after takeoff, that you will be climbing into a gradient wind that will be on your nose helping performance, not into a tailwind and its sagging shear affect. Avoid the double-jeopardy situation of poor performance from above-standard temperature and soggy airspeed struggling through shear.

✝ Engines Don't Like It Hot

Heat affects the engine's running, too—piston or jet. It doesn't put out as much power because it is flying, in effect, at a "higher" altitude.

Running a piston engine on the ground can heat up heads and other engine parts excessively. Tightly cowled engines need a lot of cooling airflow, and when we are sitting on the ground in hot weather, the engine isn't getting the flow it may need. We should keep ground-running to a minimum, and if we must sit there with the engine running, be certain our nose is headed into the wind to gather all the air possible.

Vapor locks occur quite easily in hot weather. Most engine systems are designed to prevent these; they aren't much of a worry with modern aircraft, but if fuel booster pumps are required to be on for takeoff, it's worse to forget them on hot days when vapor locks could occur. If we use that checklist, this will not be a problem.

All this suggests that cold days are wonderful. They are, for airplane and engine performance. The airplane leaps off the ground and it climbs like a homesick angel; it's one compensation for freezing.

But cold does create a hazard in relation to air density. It affects the altimeter. In very cold air, the altimeter will read higher than the airplane's actual altitude.

For example, flying at 10,000 feet with a temperature of minus 32 degrees C (minus 25 degrees F), which is 27 degrees C (49 degrees F) below standard, if the altimeter setting was furnished by a sea-level station the altimeter will say 10,000 feet, but our actual altitude will only be 9,000 feet! So if we were poking around mountains counting on 1,000-foot clearance, we might not have it.

Imagine a low approach to a sea-level airport with a very substandard temperature of minus 26 degrees C (minus 15 degrees F), such as one can get in the north. When our altimeter said 300 feet, we'd be at only *258 feet*. It's not a big difference, but big enough when you're making an approach to a 300-foot-minimum airport.

A good rule to remember is that for every 11 degrees C (20 degrees F) that the temperature varies from standard, there is a 4 perent error in the altitude: indicated on the altimeter (lower temperature, lower altitude).

So cold temperatures are fun to fly in; but they can give us erroneous altitude information, and that's disastrous. What this all says is that temperature, like wind, is one of the elements that a good pilot becomes aware of, constantly checking what the current temperature is and how it's going to affect the flight.

10

Some Psychology of Weather Flying

Before getting into weather flying, we should talk about emotions. Basically, airplanes get us into places and situations that humans weren't designed for. The inside of a thunderstorm, for example, where the airplane is battered by turbulence, where there are deafening, rapid-fire staccato noises caused by heavy rain, and sometimes hail beating on the airplane and our nerves. Lightning flashes, thunder is heard, and now and then a flash of lightning leaves an odor like the burning insulation of an electric wire that spooks our inner senses.

In clouds, with ice collecting on the airplane and a very low ceiling at our destination, we can feel pretty lonely. With a little imagination, a pilot may wonder if that cozy world is still down there at all.

Flying weather at times requires a firm grip on one's emotions. Emotions have to be subdued by realizing that to lick weather, one needs logical, step-by-step thinking and intelligent use of equipment. Wishing will not make a successful descent through the clouds and landing at the airport. A good job of flying will.

Sometimes a pilot has to take a firm grip on emotions, forcing oneself to control a nervous stomach and shaking hand. This isn't always easy, and anyone who says they've never been scared flying in weather either isn't telling the truth or hasn't been there!

What we have to do is be big, grown-up people and keep everything under control. One can be anything from slightly to completely scared and still think; and think we must.

A difficult factor is that sometimes in flying we can be nervous or scared for a long period of time. We may sit on instruments a number of hours with everything under us below landing limits and our destination on the ragged edge, too. This type of operation is known as "sweating it out," and a more perfect description was never thought up for anything.

✝ Self-Discipline

When everything inside us is scared, we have to work harder to do a good precise job of flying, thinking rationally all the time. In this situation a pilot must do the utmost to be relaxed. Being relaxed creates better flying and better thinking and reduces fatigue. So even if hell's fire and brimstone are all around, we must keep reminding ourselves to sit back comfortably, relax those white knuckles on the control wheel, and think! It isn't easy, but it's possible; and working on it, forcing oneself, and practicing make it possible. Surprisingly enough, this effort can develop a control over our emotions that favorably affects the nonflying part of our lives, from driving automobiles with their dangers to personal problems of finance or heart.

There's a natural tendency for fright to speed things up. If a person is in a jam in the air, the basic human desire is to be back on terra firma. If panic takes over, there may follow a desperate, too-fast attempt to get on the ground almost any way that seems possible. This is dangerous, and its prognosis doomful. This is when people try to go lower when they should be going higher; it's when they try to auger down through a too low and small hole, or when they try to land in a field that hasn't been looked over carefully, nor the trees, wires, and what have you around it.

What we must do when fright takes over and panic begins to spread is get up where all obstructions are well cleared or, if VFR, stay where it's safe, even in a circle, and then, in both cases, take time to think the situation through carefully, get settled down, and after that to act calmly and precisely.

While it isn't a part of the emotions of weather flying, it is related to being in trouble, and so we'll talk about asking for help. A pilot shouldn't be bashful or reluctant to say that he or she is in trouble or getting into trouble. The radio is there to aid pilots. There's plenty of help around, and a good, clear yell will bring it running. Say what's wrong, where you are, what's needed most. Use the frequency currently being used or the emergency frequency, 121.5 MHz. The calmer one can be while doing this, the easier it will be for someone to help.

But it is well to remember; asking for help should be in the category of last resort and not a crutch used routinely to help fly weather. Flight Watch is fine, talking to an FSS is fine, too, but don't think ATC is there primarily to get you through thunderstorms or other bad weather. Their job is traffic separation, and all other services take second place except, of course, in emergencies. But emergencies should be emergencies, and not thought of as "routine" help.

✛ Think, for Real

People rarely get into the situations we are talking about here, and weather flying isn't a desperate thing at all. But one reason people do get into trouble is that they don't do enough logical thinking in advance. The preflight analysis of weather, the decision to go or not, the planning of fuel, and the cold decision of whether the airplane and pilot are adequate must be done logically and objectively. Do not take on weather beyond the airplane's capability. A Cessna 172 cannot fly weather that an equipped Cessna 210 can, or a twin, and on up. Simply, and briefly, we need range; enough power to climb fast and high in certain conditions: deicing; radar or spherics; instruments, radios, and autopilot to relieve the multiple workload that weather and the ATC system demands. Hunches don't have any place here—with the exception that negative ones cannot hurt, such as, "I've got a hunch that place will fold up, so I'm going to take more fuel." Actually, that probably wasn't a hunch, but a judgment made after studying the weather.

But irrational optimism, like thinking, "It's going to be okay," on the basis of a keen desire to get where one wants to go, is taboo in this business. If a pilot uses logical thought processes and keeps emotions under control, it will become possible to handle tough problems with composure. Being prepared with proper briefing, keeping up with weather's changes, knowing equipment, and feeling competent with practiced flying, one finds that desperate situations seldom seem to occur.

11

Turbulence and Flying It

Turbulence comes in all sizes, from a little choppiness to big, hard clouts. It affects us near the ground and up high where jets fly; even U2s at 70,000 feet find turbulence. Unfortunately, we cannot see the motions of air, and there is still doubt about how, exactly, the air moves when it's disturbed. Flying gliders close to mountains has taught me much about the interaction between wind and mountains that creates turbulence, sometimes very nasty turbulence. And working a glider in mountainous areas, I've learned about waves and how to avoid or benefit from their fascinating action.

Flying over the sea, too, one cannot look down on the endless waves, whitecaps, and swells without thinking that the ocean below probably isn't much different from the one we fly in. They just have different densities. A wave comes toward the shore and hits a line of rocks. The water sprays and tumbles over the rocks, creating a confusion of wild water. Make the rocks a mountain and the water fast-moving air (wind), and the similarity becomes obvious.

Is this much different from wind hitting the trees on the approach end of a runway, being thrown up and broken to descend in a confused manner toward the ground downwind of the trees? No, I don't think it is.

And way up at 35,000 feet, couldn't turbulence be a moving mass of air in waves? The wave breaks and, if we could see it, might look like an ocean wave, the kind surfers ride.

All this comparison may be amusing to think about, but how does it help? It helps in visualization. Our wave breaking over rocks, which would give us a wild time riding in a boat, can be much like the wind coming against obstructions; it will be disorganized, so that our airplane will roll, yaw, and pitch, and it will sink. We must be prepared to combat this.

Visualize water rushing down a stream bed and then flowing over smooth, downward-sloping rocks. It follows the contour of the rock bed. A runway perched up high, with the ground sloping away from the approach end, is the same thing. Coming in to land, we fly over the sloping area, where the wind flows downward, and we sink as we fly into this downward-flowing air. A marginally low airspeed, plus aiming to land short, may suddenly find us touching down before the runway with a good chance of leaving the landing gear back on the slope as we slide on our belly along the runway.

What we are saying is that a good pilot visualizes the air motion around an airport, a mountain, or any obstruction near the flight path. As we fly by an airport on downwind leg, we should have in mind the wind direction, velocity, and gustiness. Then, with crafty eye, we look at the obstructions and visualize how the air is moving over them and how it will affect our airplane. We should be prepared for choppy air, a downdraft, or sometimes an updraft that will make us balloon when we don't want to.

Be aware that the wind direction is not always directly down a runway, but often across it to some degree or other. If obstructions, such as trees or buildings are to the side of the approach path to the runway, then any cross component of the wind will cause spillage over those obstructions, making the approach a turbulent flight, frequently with strong downdrafts.

Mountains can cause up and down drafts quite a distance from an airport, and one should look at the terrain for many miles in all directions and try to visualize its influence.

I experienced a good example of this when taking off, with my son, from Honolulu in a Cessna 402 for Tarawa, 2,400 miles away. It was a ferry flight and we were overloaded with extra fuel. Once in the air, after a long run, we made a right turn and headed out to sea. Our climb was very poor, but after flying a few miles away from the island, it suddenly increased to the normal rate for the load we had. This was to be expected because the ground slopes up to about 2,800 feet six miles northeast of the Honolulu airport. The pleasant northeast trades were flowing down the slope of this Koolau Range, as it's called, causing a large area of settling air we were trying to climb in. It kept us down until we flew out of it eight to ten miles from the mountains.

Some basic—but important!—views of wind effects around an airport.

✈ Kinds of Turbulence

Turbulence can be categorized according to its location. These are the places and areas pilots find turbulence: near the ground; in those obstructions we talked about; in the convective layer of the atmosphere or, in simpler words, in the haze below the inversion; in clouds, of course, and the more unstable the cloud, the rougher; and finally in clear air. This last is not to be confused with the turbulence in the convective level. It is turbulence in clear air well above the convective level, generally in connection with a jet stream. But there are other things that may cause it, and we'll go into them later.

✈ How We Fly Turbulence

We should talk about the ways we fly an airplane in turbulence. Basically, it's a matter of not fighting, but rather letting the airplane have its way with the little displacements it wants to make. We don't sit there and madly fight the stick or wheel.

This, of course, has its limits, and there comes a time when we say, "Whoa, baby, you've gone far enough." Then we move controls and make the airplane come back where we want it.

Since an airplane pitches, rolls, and yaws, we might look at turbulence from those aspects. Let's start with pitch, the up and down movement of the nose. A gust will make the airplane's nose change position. What we want to do is keep the nose where it should be, and that position is the place on the horizon, real or artificial, for the speed we are trimmed. We fly this attitude and keep the nose near that point on the horizon.

If we take off or put on power, this position will change, and we must retrim. But these trim changes are small. What we want to avoid is large displacements of the airplane. In clear air on a calm day, try to see what positions the nose takes for different speeds and with different power settings. One quickly learns that big power changes make big attitude changes. This is a clue, of course, that juggling power during turbulence in large amounts will result in big attitude changes, out of trim control forces, and general confusion in flying. We don't want that. It's best to know the speed one wants, trim for it, and then ride it out, keeping the nose very near this trimmed position.

On a clear day one should try and then memorize pitch settings on the artificial horizon versus power settings and trim, for everything from best turbulence penetration speed to climb after takeoff. They will vary with load and altitude, but will be valuable as something to grab quickly when needed—at least as a starter. This knowledge is also useful in the event of airspeed indicator failure, iced-up pitot, taped-over static port, or mechanical failure.

Altitude will vary. If it's very rough and you are worried about the structure, let it vary! Otherwise, make the smallest possible power adjustments and keep the flying technique as simple as possible.

Roll and yaw are mixed together to different degrees in different airplanes. Push a rudder pedal and the airplane yaws; as it yaws, it will begin to bank, sloppily, but it will. In a swept-wing jet, this tendency is very strong. Push the rudder and immediately it starts to bank, too. Bank the airplane, on the other hand, without moving the rudder, and its nose will eventually start to yaw around.

Another thing that happens with bank is that the nose goes down. Bank the airplane, and the nose starts down; unbank it, and the nose rises. The important time, however, is when it goes down. That's how spiral dives begin if a pilot is on instruments in rough air and is not experienced. The airplane banks, and the nose wants to go down. If the pilot pulls back without doing anything about correcting the bank, the bank steepens, the nose drops more, and the speed goes up. It's a spiral dive off and running! It happens very fast. Many airplanes in this condition will, when improperly flown or not corrected, reach red-line airspeed in something like five seconds or less!

So, keep the wings level! It's simple, really, to keep the wings level and keep the nose where it should be on the horizon. Do it in a relaxed way with easy pushes and pulls, not jerks and shoves.

✝ Convective-layer Turbulence

Now let's talk about turbulence in the convective layer, the stuff below the haze level. Today's a day for it. A front passed yesterday, and from my study window, across the rolling country, the visibility is good in a brisk northwest wind. There are small, widely scattered cumulus drifting rather fast across the sky.

Just a glance at all this says it will be rough from takeoff until I get on top of the CU—probably only about 5,000 to 6,000 feet because it's November, and in colder weather the tops are lower than in summer.

There are mountains in my country with farms, fields, and woods. As we take off and climb, it will be rough. Turbulence will start pushing us sideways as soon as we leave the ground, jolting and jarring. That will last until we reach 800 to 1,000 feet because of the turbulence of the unstable air being joined by the turbulence of that strongish wind bouncing over the hills, trees, and fields. So there are two types of turbulence: orographic turbulence caused by terrain, and instability turbulence within the air mass.

Above 1,000 feet it will still be rough, quite rough, but some of the small, jiggling, jolting, fox-terrier kind of action will be gone as we get up out of the orographic influence.

We reach the CU level, climb up above that, and suddenly, like flipping a switch, it is smooth. The haze is below us, and at our level, visibility is truly unlimited. It's the place to be.

But let's go back and wallow around in the rough air below. How rough it is depends on how unstable the air is and how strong the wind is. The fresher the air mass (that is, the first day or two behind a front), the stronger the turbulence. The air is colder than the ground, which heats it and sends it upward in fast-climbing thermals. These are the days glider pilots become giddy. The sky is full of lift, and they make records and go long distances.

But in an airplane, each thermal we go through is a bump. Some of them are pretty strong bumps. Out West it's not uncommon to fly a glider in thermals going up more than 1,200 feet per minute. Hit that column of rising air at airplane cruising speed, and it's a good jolt.

✛ It's Rougher Than You Think

I've often thought, while bouncing along in the convective layer, that if I were on instruments, near a thunderstorm, hitting bumps this hard, no stronger, I'd be quite concerned, slowed down, and alert to use best turbulence flying technique. But because it's clear and sunny, we don't think much about it. Sometimes we should because the airplane's structure is taking a beating.

One special time is during descent. We often see a pilot push the nose down, for descent, and let the airspeed increase right up until the airspeed needle is tickling the red line. It's smooth, and the airplane is making impressive time. But it descends through the inversion into the convective layer and suddenly is hitting really solid bumps. The combination of very high speed and the strong bumps is certainly putting a heavy load on the structure.

Fly above this kind of cloud. Beneath these cumulus it will be rough and uncomfortable, while on top the air will be clear, smooth, and cool. (NOAA PHOTO)

An alert pilot will reduce airspeed before whamming into the roughness below the haze level and then, after the bump's intensity has been felt, make a decisions as to the best speed to use the rest of the way down for comfort and the airplane's structural integrity.

There's little reason to stay down in the convective layer and bounce around if you can top it. In the East this is generally fairly easy, because the top of the layer isn't much over 7,000 in winter and 10,000 in summer. In the Far West it's another matter. Because of the high ground and strong heating from the semiarid land, the haze level will often be above 15,000 feet. Then you need oxygen or pressurization.

Flying in the constantly bouncing air is fatiguing, and for passengers it's uncomfortable. The bouncing doesn't help the aircraft structure either. This isn't a serious point, but it is worth considering.

✝ Dust Devils

On windy, rough days, especially in the West, dust devils form. These look like miniature tornadoes that work across the countryside kicking up dust in a swirling cloud. They are easy to avoid, and it's a good idea to do so. On landing one could get roughed up near the ground and settle drastically or hook a wing as one is tossed around. The action feels much like prop wash or jet vortex; it has that uncontrollable feeling. Even after one is on the ground, the dust devil can lift a wing or flip an airplane over.

Big dust devils are obvious, but smaller ones may only look like a gossamer, funnel-shaped ghost skittering along. If we weren't looking for it carefully, we might fly into it.

When dust devils are forming, it's important to have airplanes on the ground well tied down. A dust devil can turn over a parked airplane easily.

✝ Turbulence Near Mountains

Flying during strong wind conditions in mountainous areas, particularly when the air is unstable, can result in some very rough rides. The windward side of a mountain range will have predominantly rising air. It will not be consistent, but rather choppy and gusty. The gustiness will show itself in great ballooning climbs that will make you feel you'd like to reduce power, followed by sinking feelings as the up-gust diminishes. Often this variation will simply be the result of flying from one mountain contour to another. Mountain ranges are rarely ruler straight. They have cuts, and the range twists so that its face is at different angles to the wind; the mountain's slope angle will differ from place to place, too. All these irregularities create variations in the winds over the mountain. Glider pilots who have done a lot of ridge soaring learn this, and an experienced one can look along a ridge and tell what kind of lift and turbulence will be found and precisely where.

On the downwind side of a mountain, the air will mostly be going down. It will be chopped up, spilling, and rough. The place to fly is well away from the mountainside. In unstable air, a mountain—especially the upwind and sunny side—will produce thermals, rising shafts of air, that add to the confusion of turbulence.

We don't always fly along mountains, but often we cross them at right angles, but never approach at a 90-degree angle, make it 45 degrees, then if you're having trouble getting over it's a shorter turn to get back to the valley. When we approach a ridge, it's important to be aware of the wind. Is it a tailwind or headwind?

Say we are flying low, 500 feet or so above the ridge tops. With a tailwind, the airplane will climb as we approach each ridge. We are being "zoomed" by the air flowing up the slope. But when we pass the mountain ridge and fly to the downwind side, the airplane will sink, and we'd better be prepared for it.

Flying the other way, toward ridges into a headwind, is hazardous because approaching the ridge, we enter its downwind side. We could also call this the down slope side. We will sink, often below the mountaintop, and need power to keep from getting too low and too slow. We shouldn't just pull the nose up and let airspeed drop while trying to climb. Use lots of power and get up, or turn away from the condition with sufficient airspeed; run from it! But it is best never to fly low near mountains, especially the downwind side, particularly if the wind is strong; we should be particularly careful flying toward them into a headwind, since one can sink and get in a dangerous position. Never approach a mountain ridge from the downwind side except with lots of altitude; I mean 2,000 feet above the mountain—more if it's a big mountain and the wind is strong.

A hazardous place to fly is between two ridges so closely spaced that downward-flowing air from one ridge is turning to climb up the next ridge. The air movements will be confused, and there will be lots of sink.

When flying along or across a mountain range, realize that the wind speed will increase where cuts and passes go through the range, because of venturilike action, and the turbulence will become worse.

The wind speed over the top of a mountain increases in much the same way that air flow increases over the curve of a wing. In the extreme, Mt. Washington in New Hampshire has recorded winds over 200 mph! It wasn't the general wind going that fast, but rather the wind accelerated by the curve of the mountaintop. Almost any mountain will have this effect to some degree. Where the wind speed increases, there will be shear and turbulence on the top and the downwind side.

Of course, what all this means is that we shouldn't fly close to mountains when the wind is strong and the air unstable, especially the downwind side. If you are close to a ridge, never, never make a turn toward the mountain. Always turn away. To really respect and learn about flying in mountains, I strongly recommend ridge flying in a glider...it's fun, too.

✝ Mountain Waves

Mountains create another type of turbulence that extends above the convection level; it's connected with mountain waves, sometimes called *lee waves* or *standing waves.*

Fast-moving air hitting the side of a mountain causes waves in the air downstream of the mountain. They are like waves in a river behind a submerged rock. Waves in the sky sometimes extend to great heights. The front side of the wave is going up, and glider pilots like to hunt for this area and make altitude flights. Paul Bikel broke the world altitude record in a glider by going to 46,267 feet in a wave off the Sierra Nevada of California! From 4,000-foot mountains in Vermont, we get as high as 29,000 feet. I've felt wave influence at 7,500 feet from a 700-foot hill.

Let's visualize these waves. They form downwind from the mountain, not very far off, in fact almost over the peak of the mountain. The waves repeat downwind with second, third, and fourth waves and often lots more. The waves go up on one side and down on the other. We often discover them by noticing the airplane wanting to go up. We push the nose down and get increased airspeed without

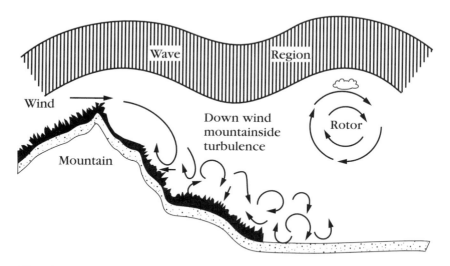

This shows two types of mountain turbulence: one is the tumbling air in the lee of a mountain; the second is the rotor under a wave. The little CU on top of the rotor, a small shredded wisplike cloud forming, is the cue telling where the air is very rough. The two types of turbulence can occur at the same time or separately when there is no wave action but lots of wind spilling over a mountain.

adding power. We are in the up part of a wave. The air is generally glassy smooth.

On the other side we get a smooth down flow. We have to pull the nose up and add power to hold altitude. This area is smooth, too. So far, so good.

But under the wave we find turbulence in an area called the rotor, where the wave air conflicts with the undisturbed normal air below and causes a tumbling, rolling, chopped-up, confused mess. It is rough, how rough depends on the strength of the wind, the mountain, and how much of a wave there is. It can be very rough! In severe cases this rotor has torn an airplane apart. It's not a thing to fool with.

How do we know it's there? First, with strong wind flow across a mountain, especially after a front has passed, we can expect wave action. Looking up into the sky, we may see lenticular clouds. These are a characteristic sign of waves. They are long, slim, lens-shaped clouds, but the real key is that they do not move as normal clouds do. The reason for this is that the rising side of the wave cools air to its condensation temperature and forms the cloud. As the air starts down the backside of the wave, the pressure increases, the air gets warmer, and the cloud disappears. So the lenticular cloud forms and disappears on the wave, staying with it, and doesn't drift along with the wind. It is very turbulent under lenticulars or where they would be if enough moisture were present to form the cloud. Frequently there is not enough moisture to form a cloud, so there can be wave with no lenticulars. We should not be lulled into thinking there is no wave because there are no lenticulars! Any time the wind direction is across a mountain range—or within an arc of about 70 degrees of perpendicular, and the wind maintains this direction with height, and the velocity increases with height, be suspicious of wave action.

In the valleys downwind of the mountains, the rotor often extends almost to the ground; the air near the ground is rough, and one can see clouds of dust swirling into the sky, if it's a dry, dusty area. I've flown through a rotor in Vermont that was almost on the airport at Warren. Descending to land, it was wildly turbulent.

Flying downwind of mountains, we want to be especially careful when the possibility of a wave exists. Sometimes we can get the combination of air-mass instability and wave-rotor action together, and it makes a mighty interesting ride!

Rotors can lie fairly high in the wave structure, but the most turbulent rotor will be found at the mountain level and below. If the mountain is 4,000 feet high, then the rotor will be roughest from

Beautiful lenticulars that tell a wave is working in the sky. They look smooth and peaceful, but below it can be very turbulent.
(NOAA PHOTO)

Lenticular clouds from above.

Action on the downwind side of the mountain range. In the clear air immediately to the left of the mountains, there is a downdraft—a very strong one that only powerful airplanes could handle, and the pilots would have their hands full. Where the dust is rising the air is moving up in strong wave action and is very turbulent. It would be best to avoid the entire area. (PHOTO BY ROBERT SYMONDS FROM DE VER COLSON COLLECTION)

4,000 feet down. To be on the safe side we should add 2,000 feet or so to the mountaintop altitude to assure being out of the worst part of the rotor. If you can see the rotor clouds, fly above them. For bigger mountains, it would be advisable to add extra altitude above the mountain height, say 5,000 feet more. But the best way is to avoid the area altogether.

Waves can be created by any size mountain, but they are not always visible as we've explained. The rotors, however, will sometimes show themselves as little pieces of shredded cumuluslike cloud. There may be moisture in the lower levels that will support, or almost support, small CU, while there is not enough aloft to make a lenticular. So if you see CU on the downwind side of a mountain, generally not much higher than the mountain, and when you study them carefully you see a shredded, moving, turning-over appearance to the top side of the cloud, look out! That's probably a rotor. Sometimes there will only be a few small, half-formed, gossamerlike wisps of cumulus that you almost "feel" rather than see, but if we're watchful and see them, they can tell an important story. Those innocent-looking wisps may signal a very rough rotor. Any atmosphere like that downwind of a mountain should be cause for suspicion and, if one must fly through, it is time to prepare for flight in turbulence.

Any flight downwind of a mountain, even a small ridge only 500 feet high, can have a nasty rotor if the wind is strong and the air unstable. Flight down low is not the thing to do. It's no time or place to be circling a friend's house or doing other foolish, low-level flying.

When getting in the uplift of a wave, the air is smooth and might fool a pilot flying an aircraft with low power and no oxygen. It's pos-

The shredded cumulus of a rotor under a wave. These are signposts announcing very turbulent air.

sible, but not likely if you are alert, to go up at a fierce rate to a high altitude where oxygen is needed. The way out of such lift is to turn downwind. But the aircraft would then get in the *down* part of the wave, which is also generally smooth. The descent rates may be high, more than one could overcome with the engine. Pulling the nose up and pouring on power is not the way to fight the down flow because it's a sucker set up to get into mushing, stalled flight, and not gain anything in the fight to combat the severe down flow. It would be best to keep up speed and fly fast downwind to get out of the wave condition. You might be going down fast while doing this, but the combination of airplane high speed and the tailwind will get you out of the down current of air quickly. The pilot might go through a number of waves in his or her downwind dash, but that's the fastest way possible.

The waves do not have a very long wavelength in miles of distance, so it's easy to turn and get out of the down and quickly fly into the up part. With our 4,000-foot mountains here in Vermont, the wavelength is seldom over a few miles wide. It would be more for larger mountains, but still not a big distance. If you're on a flight plan and encounter the down part of a wave and have trouble holding altitude, tell ATC the problem and that you're turning around. With marginal performance, never try to outclimb a wave's downside by pulling the nose up with consequent speed loss; stall is never far away unless the aircraft has enough power to climb and maintain good airspeed.

A special situation is a single mountain that sticks up alone, like Rainier in Washington State or Fuji in Japan. Under strong wind conditions, the downwind side can have a combination of wave and vortex. The air spills over the top and around the sides and sets up a great whirlpool action downwind that has severe turbulence. A 707 was torn apart in this condition near Mt. Fuji. So it's well to be very wary downwind of peaks that stand alone enough to create this situation; go around or pass any lone peak on the upwind side.

Mountain waves are interesting, and playing with them in a glider is not only fascinating, but a good education for the airplane pilot. I've landed many times in Milan, Italy, which is snuggled against the southern side of the Alps. Because of glider experience I'm wave-conscious and have studied and learned which conditions make the biggest waves in the area; generally, a quick look at the winds-aloft pattern tells me if I have to be extra careful while descending into Milan because of the possibility of wave penetration.

Chapter Eleven

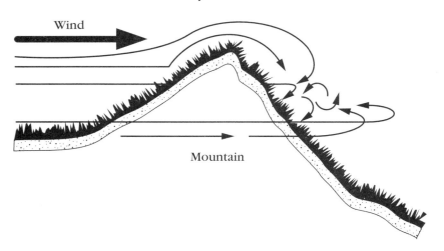

A single peak has lots of disturbed air downwind, caused by vortex around both sides and wave action over the top. Give the down side a wide berth of many miles, but it is best to go around the peak's upwind side.

An unusual and dramatic picture of a vortex situation downwind of a lone peak. If there wasn't enough moisture to form a cloud, this condition would not be visible and one might fly into it unexpectedly. Lesson: stay away from downwind of lone peaks when winds are stronger than light. (PHOTO BY JOSEPH SCAYLEA)

Incidentally, the strongest waves there come with southwest winds from the direction of the French Alps.

Wave action is something we can use to advantage on cross-country flights if we're paralleling mountain ranges. On top, particularly after a cold front passage, we feel periods of descending air, and our airspeed drops way off as we try to maintain attitude. At other times we seem to get a great boost and an increase in speed way over normal cruise. What is happening is that we're flying in and out of waves, perhaps only mild ones. If we watch the top of the clouds we'll see undulations like swells on an ocean. If we eyeball these and fly to stay on the upwind side of a "swell," we'll be in rising air and going fast. Moving our course around to fit these upflow areas may well be worth small detours and, as we said, the wavelengths aren't very large in miles, so "fishing" upwind or downwind may quickly get you out of sinking air into lifting air and an airspeed boost. At times we may feel this condition in clear air when we cannot "see" the waves. If so, do that "fishing" to find "up" air. This awareness of the air's motion may save both fuel and time. It's the stuff glider pilots work with, and it opens an exciting new concept and knowledge of the sky.

Photo taken from under primary wave looking downwind where clouds of secondary and tertiary waves are seen.

✈ Turbulence up High

Sometimes turbulence that is not CAT is felt well above the convective layer, at 12,000 feet and up. This is generally light and associated with overrunning warm air preceding a warm front. It may also be there because of other air-mass changes. Often a look up will show altocumulus clouds.

The Big High-altitude Turbulence

Now let's talk about clear air turbulence (CAT). This type has been a bogeyman in high flying for some time; it's been popular in the press and is often called the "airman's enemy," which it isn't, but it does bear watching. Let's look it over calmly.

Most clear air turbulence is associated with the jet stream. The jet stream is a hoselike band of high-speed winds at high levels. It is one manifestation of constant movement in the atmosphere as the unequal temperatures of the earth—hot tropics and cold arctic—try to balance, with warm air working north and cold air south. There are fronts high aloft that are part of this process. It is there that jet streams wander, squeezed tight under the tropopause with high winds and turbulence that influence flight as well as the movement of surface weather. Since the temperature balance isn't regular and neat, the jet stream doesn't string out exactly east and west very often. Generally, it wanders in a serpentine fashion, so if one flies east or west, one may cross it a number of times.

The jet stream is up in the high latitudes in summer, 60 degrees or so, and down south as far as 20 degrees in winter, but just to be difficult, wanders north and south during both seasons. The hoselike part is the area of maximum winds that sometimes blow 200 knots. There are also strong winds for hundreds of miles to the side of the special high-speed jet. The high-speed winds spread over a wider area to the right of the jet stream, looking downwind, than they do to the left.

Where these high-speed winds rub against lower-speed winds, there is tumbling, turbulent air. Where air mass densities change, as in a high-level front, it's rough. Where a jet stream slams into slower-moving air ahead of it, it's turbulent, too.

Meteorologists can locate the jet stream rather accurately, but they cannot pinpoint exactly where the turbulence will be. They can tell you only in general terms. After spending 8,000 hours above 30,000 feet, I can assure you that sometimes it will be rough when they say it won't and smooth when they say it'll be rough.

How rough is it? Moderate at times, perhaps severe by some people's standards. Personally, I think the danger of the turbulence itself is overrated. It's uncomfortable and at times disturbing, but it isn't anything that can't be handled.

It is possible, however, to get into trouble because of control problems. Up very high the airplane has low thrust; it's mushing along at a higher-than-normal angle of attack. It's squirrely, as the expression goes, because its Dutch roll tendencies become more pronounced. In turbulent air, almost all of which is caused by shear, airplanes can approach stall and want to go off on a wing. It takes careful flying to keep level and under control.

This isn't a desperate situation if the pilot is on top of it and realizes that it may be necessary to give up some altitude to keep control. This nervous area is why airlines try not to operate above any altitude where the airplane is close enough to stall, so that a 1.3 or 1.5 G bump might cause the airplane to exceed the stall angle of attack.

Yaw dampers are a help in these conditions. They aid the pilot and are a control augmentation or, in a small way, a fly-by-wire concept in which electronic gadgetry makes flying easier. Modern military aircraft are loaded with it; so are the modern transports such as Boeing's 777 and most AirBus airplanes. There's nothing to be ashamed of, or feel inferior about, because a gadget can fly better than we can. It's a great thing and should be used with gratitude and enthusiasm. There will be much more of it in the future.

The formula is to fly slower and fly as one would fly any turbulence, by attitude, with wings level and pitch and power changes at a minimum. Doing all these things will make CAT a nuisance and not a disaster.

✝ Where Is It?

Despite the difficulty in forecasting CAT, there are some cues the pilot can watch for to tell where CAT may be. Before takeoff the pilot should have a look at a high-level pressure chart, the 300 millibar one, which is about 30,000 feet. Its wandering isobars will show where the jet stream is apt to be. It's like any pressure chart, with the distance between isobars telling how strong the wind is: the closer the stronger.

In studying the wandering curve of the isobars, the pilot should note especially where the isobars bend and change direction. A trough, for example, will have the isobars oriented from northwest to southeast on the west side; then they will turn a corner and orient themselves southwest to northeast farther east. In the area of the corner, where the wind changes direction, you can almost count on finding CAT.

Then check the isobar pattern. Let's say the isobars are all crowded together in an east-west direction; the wind is moving fast. But farther east the wind slackens, and the isobars fan out. It will be rough in the area where the wind velocity changes. If the velocity changes faster than 40 knots in 150 miles, it's a sign there will be considerable rough air.

Jet streams above mountain wave areas have extra turbulence, and the type found in the rotor will be found at high altitude, but not with the viciousness the lower rotor has.

The Weather Service also publishes information telling wind velocity change with height. Generally, it is considered that if one sees a wind change of four knots or more per 1,000 feet, it will be a rough area. This I find hard to count on. Sometimes it seems to work, but I've seen occasions where the four knots and everything else said it was going to be rough, only to fly through and never hit a ripple. If this figure coincides with other turbulence cues, like bends in the isobars, it seems more valid. At any rate we watch these areas with suspicion.

✝ The Tropopause and CAT

Perhaps the most interesting thing to study in advance is the height of the tropopause, which tells us a lot. The tropopause is the place where the troposphere, in which we live, ends and the stratosphere begins. The tropopause is a narrow band separating two different kinds of air. The most noticeable characteristic difference between them is temperature. As we know, the temperature decreases with altitude in the troposphere and keeps right on doing so until we reach the tropopause, where the temperature stops getting colder. This, theoretically, is at about 35,000 feet in the average latitude of the United States. Above the tropopause, in the stratosphere, the temperature is considered to be about 56 degrees C below zero, but it varies considerably.

To make life interesting, the trop, as the trade calls it, wavers up and down with the passage of lows and highs under it and can actually range in altitude from the low 20,000s to the 40,000s right over the United States. Normally the trop is high over the equator, at about 55,000 feet, and low near the poles, at about 24,000 feet.

✝ The Tropopause Is Important

A couple of things happen at the tropopause. The temperature goes up or holds steady, as we said, and there is a choppiness as you go through the tropopause into the stratosphere. These things do other things. The temperature change is an inversion, just like one 1,200 feet above the ground early in the morning. The inversion makes sort of a lid, and if there's a jet stream, it will be fastest right under the tropopause. If this band of strong wind contains turbulence, it will be roughest right at the trop and for a few thousand feet below it. The depth of the region where the turbulence and wind are strongest is a point of argument, but personally I like to give turbulence at the trop a 4,000-foot berth, that is fly 4,000 feet under it. That doesn't avoid all the turbulence, but it makes it less bothersome.

Above the trop, in the stratosphere, the turbulence dies out within a very short altitude range, 1,000 feet or less. If you get *up* out of the trop, it generally becomes smoother quickly. Wind decreases, too. If one has a strong headwind, it is often possible to see it drop off dramatically as soon as one climbs into the stratosphere. If we're trying to avoid headwinds, and the temperature is low enough to operate up there, the place to fly is above the trop, and it'll probably be smooth, too.

Since jets fly between about 28,000 and 43,000 feet or so, it's simple to see that they wander in and out of the tropopause as they cross over high- and low-pressure areas. This means that our current jets generally fly at the most annoying altitudes. Every time you pass through the trop, it's bumpy, and if there's a lot of wind or turbulence, it may be very bumpy.

Nuisance number two in the trop is the temperature change. You take off for a long trip with a heavy airplane and plan to cruise at 33,000 feet. The weather charts say it's cool there, and the trop is well above you. But once in the air, ATC says to climb to 35,000 feet. As you go through 34,000 feet, there's a little choppiness, and you can see a slight haze level. Your eyes dart to the outside air temperature gauge just in time to see it go up to a "warm" 44 degrees C below zero. The airplane feels mushy in the "warm" air, and life is no longer beautiful. You are above the trop.

This is an inconvenience. You may stick it out and stagger until the weight lowers through fuel use or, if it's too difficult, you call ATC and tell them you just cannot fly up there. Then they have the problem of working out something lower for you.

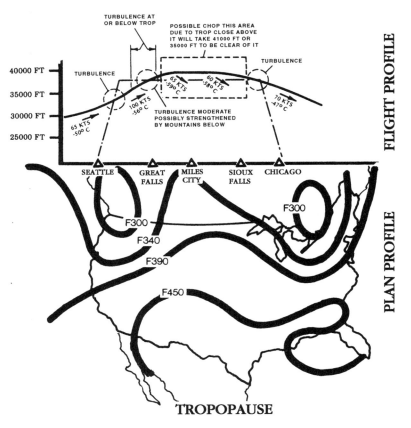

A simplified view of a tropopause chart and visualization of its effects on a Seattle-to-Chicago flight at 39,000 feet.

All this makes studying the tropopause charts very important before flight. We're looking for these things:

1. Trop heights.

2. Trop temperatures.

3. Trop wind direction and velocities.

4. Vertical shear values.

We are interested in the trop level, especially in the area of take-off and where we'll reach climb altitude. If the trop is supposed to be 33,000 feet and it's temperature −50 degrees C, we may not be able to climb higher immediately, even if we'd like to, because the temperature above the trop will be −50 degrees C or warmer. We may have to stay lower until we fly into a higher trop area, or until we burn off fuel, are lighter, and have better altitude capability.

Vertical shear will show as boxes on the trop chart with numbers in them. We might see a value such as 4. This means the wind velocity changes 4 knots each thousand feet. A value of 4 or more is generally considered enough shear to call for turbulence, but as we said, it doesn't always happen.

A trop chart has many characteristics similar to other pressure charts. If the trop levels change greatly over a small distance, the winds will be high speed and it'll be rough. In a steep slope like that, the temperatures will change quickly and, as in a lusty cold front, there will be rough air. If the trop levels make a sudden change in direction, it will be rough there, too.

The trop chart shows how high we'll have to climb to be out of rough air and strong winds. If we can climb that high, all this is very pleasant to know.

For high flying the trop chart is as important as the surface chart in telling weather. Near, in, or above the trop is where high-flying aircraft spend their time, and familiarity with the phenomena of the trop is of utmost importance.

If we don't get to a place where we can see trop charts, we should ask the briefer we talk to via telephone, or other means, questions that will tell what the trop charts show.

Pilots should not be surprised by sudden turbulence if they have studied the weather carefully, and especially the trop charts and high-level pressure charts. After that, close observation in flight will show how the forecast charts are working out and how to modify one's cautious flight if they aren't. Close monitoring of outside air temperature in flight is another clue to changes about to occur. There's little excuse for "surprise" turbulence if a pilot has studied the weather before flight and then monitors it closely during flight.

✛　Shear

Shear is an effect created by winds of different velocities and the cause of almost all turbulence. We feel it when flying from one strength wind to another. Flying from a strong headwind into a lighter wind, the airplane loses airspeed and must get it back. At cruising speed this isn't much of a problem; but at low speeds during an approach to landing it can be significant.

At landing-approach speed, we are not too far from stall. An abrupt airspeed loss, caused by flying out of a headwind, suddenly brings us much closer to the stall. We sink. If we don't do anything about this, it will take a long time to get the speed back, many seconds, even a half-minute or more. In that time we might get too low and hit short of the runway.

When this speed loss occurs, we have to get on it right away. We pull the nose up to keep from sinking and put on more power to overcome the drag from the pull up. And don't be bashful about getting lots of power on quickly.

The pull up can be quite severe in an extreme condition, and it may be necessary to pull up almost to the stall angle of attack to get up and away from the ground. If the wind shear is strong, the airplane's angle of attack will be low and it will be necessary to change that to a big angle in order to get the necessary lift. The pitch angle, what you see on the horizon, be it true or artificial, will be quite large and it may look pretty desperate, but the actual angle of attack, the relative wind, will not be that extreme. During all this, of course, maximum power must be used, with throttles right up to the limit!

The task is not to exceed the stall angle of attack, and one will have to rely on one's sense of feel and the stall-warning device on the airplane. Now there are instruments available specifically designed to help fly through shear, but it is important to note that certain shear and vortex turbulence near a thunderstorm cannot be flown. The airplane cannot handle it, special instruments or not! The only positive, safe answer is not to attempt landing with a thunderstorm on or near the airport!

If strong shear is encountered and large attitude power changes are used to combat it, a serious hazard then arises as the shear stabilizes and the pilot attempts to continue the landing. The airplane is trimmed for a steep pitch attitude and excess power. To retrim, get the power off the right amount, and landing is very difficult flying, especially in airplanes that require large stabilizer travel. Shear also often has a crosswind component which will quickly push the airplane away from the runway center line. So the pilot not only has to do some difficult trim, power, speed, and rate-of-descent juggling, but also has to turn the airplane and get it lined up with the runway, generally at low altitude. All this is very difficult flying and probably where shear accidents are actually caused. So if a pilot is fortunate enough to recover from the shear, it's no time to press one's luck and try to get lined up and land. The smart maneuver is a go-around!

Shear is easier to cope with in a propeller airplane than in a jet. The propeller airplane, when we pour on power, gives thrust quickly;

it also produces an immediate increase in airflow over the wing from the propeller wash and therefore adds lift almost immediately. The jet engine provides thrust only, which, if the throttle has been pulled full back and the engine spun down, will take a long time coming back to full power, so it takes longer to regain speed. While all pilots have to be alert for shear, the jet pilot has to be particularly so.

✛ Where Is Shear?

Shear happens in various ways. Wind velocities change as we descend (winds are usually stronger aloft and weaker near the ground), producing shear. Shear also occurs because of shifts in wind direction. In thunderstorms, when a front is approaching an airport, we can easily get a wind change from southerly to northerly in an instant that will make landing an exciting adventure.

More subtle, however, is a wind shift as we descend that changes a headwind into a crosswind and effectively makes the airplane's velocity lower.

A few hundred feet above the ground, the wind follows the isobars. That's *gradient wind*. But the surface friction of the wind running over the ground causes the wind to flow more toward the low pressure, so the wind near the ground, instead of being easterly as the gradient wind might be, will swing toward the northeast and north as we descend. This, of course, cuts down its effective velocity and so produces a shear effect with an airspeed loss.

It's a good idea, when landing in a crosswind, to be conscious of what the gradient winds are. Is the wind aloft more of a tailwind in the direction we're landing, or a headwind?

If it's a tailwind, we may overshoot or land too long. For example, we are landing to the west. There's a north wind on the ground. The wind aloft is east at about 40 knots at 1,000 feet. We are making our approach with a whopping tailwind. Our ground speed is high. As we descend, the tailwind decreases, but we only have a short period in which to get rid of a lot of ground speed before we get to the runway. We tend to overshoot, and so we push the nose over to get down to the runway's end. Our speed increases, and lift becomes excessive. It's difficult to touch down on the runway; it would be wise to abandon the approach.

So we pull up and try landing to the east. Now we have a headwind aloft, but as we descend we lose the headwind and some airspeed. Now we'll have to correct for sink and overcome the tendency to get below the glide path.

It's obviously important to consider what the gradient wind is as we let down to land. It will give us an idea of what to expect during the approach. Will we be scampering and diving to get down and land within the airport—a tailwind condition? Or will we be pouring on power and lifting ourselves up to keep from undershooting—a headwind condition—when the wind decreases as we descend?

Sometimes the difference between the gradient wind and the surface wind can be quite high, even more so at night than in daytime. A nighttime inversion makes things seem still on the surface, but just above the inversion the wind can be swishing along at 50 knots if there is a strong pressure gradient. We must therefore be wary of large differences between surface winds and winds aloft when the gradient is tight and there is a strong inversion.

Warm fronts can create strong shear conditions when things on the ground seem tranquil. First, let's review the wind directions on both sides of a warm front as well as above and below the frontal surface which slopes, from the front, generally toward the north, about 300 to 1, which says that 300 miles north of the surface front the front aloft will be at 5,280 feet above the ground. The winds, north of the front and below the frontal surface, are north and easterly, often rather light. South of the front, and above the frontal surface, they are south to southwesterly and often quite strong.

Now let's place a warm front 30 miles south of Kansas City, where we will land. The surface wind at the airport is a gentle 10 kts or less and northerly, but above 600 feet the wind is southwest at 50 kts! So as we descend toward 600 feet we have a tailwind of 50 kts. Our rate of descent will be high and so will our ground speed. We're all set for an overshoot. This isn't just hypothesis. It has actually occurred at Kansas City.

A warm front is not especially bad because we may have to go around, but it becomes serious when the pilot doesn't go around and tries to make a landing, touches way up the runway, and goes off the end into who knows what!

Overshoots also may put us in a position to undershoot! Descending fast, trying to get down to the glideslope, we pull off all power. Then we run out of that strong tailwind. Our ground speed slows, but airspeed increases and we go above glideslope. In an effort to get back down, our power is still reduced. Then, sooner than we realize, the airplane goes below glideslope and we are undershooting! Now it's pull the nose up and pour on power. Quite a juggling act, especially on instruments. So winds only a short distance above the earth can be very different from winds on the surface. Tailwind-to-headwind, wind reducing in descent, wind increasing in descent, or headwind-to-tail-wind all require different recovery techniques and aren't restricted to warm fronts, but can be experienced with cold fronts, nearby thunderstorms, mountain downdrafts, and anything that can give the wind different vectors and velocities.

All this theoretical talk isn't much good if we don't know where fronts are or if we suddenly run into shear conditions and don't know why they are there. Then what?

First let's recall the two basic things about shear: A headwind on landing tends to make us land short when wind velocity decreases as we descend a tailwind tends to make us overshoot. Those are the basics.

So what we want to know is how strong the conditions are and how much the wind will change during descent. In fancier language, what is the wind gradient? This is part of our weather briefing—to give thought to wind changes with altitude near the ground at takeoff and landing airports.

Once in the air, the sophisticated way to know wind gradient is by having an inertial reference system (IRS) which tells ground speed. If the ground speed is lower than airspeed, which is a headwind, and the surface wind is not very strong, there's a steep wind gradient and we'd better be ready for that undershoot situation when we fly out of the headwind down lower. A high ground speed means a tailwind and an overshoot. An IRS can give actual wind direction and velocity so the pilot can compare it with the latest reported surface wind and know how much change there will be in descent.

All that's fine, but most airplanes don't have an IRS, so what about the rest of us? If we're on instruments coming down on ILS, it's fairly easy to get an idea of what's going on by our rate of descent.

Let's say our approach speed is 120 kts. To make good a 3-degree glideslope, without wind, we'll descend about 600 feet per minute. So if we find that it only requires 400 fpm (feet per minute) to stay on the glideslope, we've got about a 40-kt headwind. If the surface wind is 10 kts, then somewhere during descent we're going to lose 30 kts and have to get power on and scramble to keep from getting too low. If our necessary descent rate to stay on glide is near 800 fpm, then we've got a 40-kt tailwind and it's going to be a scramble not to overshoot.

Required descent rates versus ground speed often are listed on the bottom of approach plates, but if they aren't, there's a formula to help decide what the ground speed should be. Half the ground speed times 10 equals the descent rate, or nearly so. Take our 120 kts; half that is 60, times 10 equals 600 feet per minute. This is 36 fpm short, but is close enough. This, of course, is for a normal 3-degree glideslope.

We can do it backwards: if we're coming down the glideslope at, say, 900 fpm, we can divide that by 10, which equals 90, and multiply that by 2, which comes out 180, and that's our ground speed. If our normal speed is 120 kts, we've got a 60-kt tailwind!

These numbers are for true airspeed, which if we're near sea level and the temperatures aren't wildly off standard will be pretty good. If we're landing at a 5,000-foot-above-sea-level airport, however, we'll have to compute what the TAS is for our 120-kt IAS approach speed.

The formula, put up somewhere handy in the cockpit, will also be useful to help preguess the descent rate you'll have to use coming down the ILS when you know what descent winds will be. For one's particular airplane, with which we always use a certain approach speed, it's possible to mark the appropriate ground speeds opposite the appropriate descent rates next to the vertical speed indicator.

A visual approach is more difficult because there isn't any exact glideslope guidance unless there's a visual approach slope indicator (VASI) or PAPI, Precision Approach Path Indicator, and other types, which you can pick up well out from the runway. Staying on a VASI glide path, and seeing what rate of descent is needed to do it, is the same as coming down an ILS glide path of 3 degrees, if it's a 3-degree VASI—and most of them are.

But an all-visual approach, without a reference to help, means getting back to basics. If we can pick a spot on the ground we want to land on, and then watch it in relation to our descent path, we know how we are doing. If the spot climbs in the windshield, we're undershooting, and if it descends we're overshooting. So it's a sort of visual glide path. If we notice it's taking lots of power and a higher-than-normal nose attitude, we've got a headwind. If the nose is down, however, and we keep wanting to pull off power, we have a tailwind.

It isn't always possible to pick a spot in poor visibility or way out from the airport. If it isn't, then it's eyeballing the situation and using one's good seat-of-the-pants feeling and visual cues that come naturally. If we're dragging in, there's a headwind, and vice versa.

Another type of shear is caused by gusty winds. It's obvious that a wind blowing 10 knots and gusting to 20 can give one a very quick airspeed change of 10 knots. That's why, on gusty days, it is wise to carry extra airspeed to take care of these sinkings of the airplane on approach as a gust dies down and leaves one low on airspeed. Most people take their normal approach speed and add half the gust velocity as a cushion during gusty conditions, plus another 5 knots if the wind is above 15 or 20 knots.

Shear is often overlooked as a takeoff hazard, but it's as important then as in landing. There have been cases where a takeoff was made into a headwind with a strong opposite wind (tailwind) whistling along just above the ground and resulted in an accident.

If there is a front of any sort near the departure point, one should discuss the possibility of a sudden wind shift with the meteorological briefer; if a briefer isn't handy but a front or thunderstorm is, we should be prepared for a wind direction change after takeoff. A thunderstorm, realistically, is the weather most likely to be violent enough to make the dramatic wind shift that causes trouble. I say it again: a thunderstorm near the airport, landing or taking off, is a very real danger, and sometimes one that cannot be flown successfully! To say it as emphatically as possible, do not land or take off close to thunderstorms. Our present use of radar, both in the airplane and by ATC, tends to have airplanes cutting in too close to thunderstorms! Often a mild-looking storm just off the approach path can suddenly, in moments, reach the point where it dumps water and wind in violent fashion. Thunderstorms are not static; they are constantly changing. They are something to give wide berth; the accidents that have occurred on approach to landing and takeoff are gruesome enough proof.

What do we do about all this? If the front or thunderstorm is close and strong, let's wait until it passes. If we get caught because we didn't think it was that bad, then consider that a loss of 50 kts, or a severe downdraft, is going to require an excess of speed as soon as possible after getting airborne. We want to get that excess quickly, remembering, of course, not to fly back into the ground or get too low on instruments or at night, in the process.

Sudden airspeed changes due to shear, bumps, or whatever affect the aircraft's pitch attitudes. Two points: When getting a positive airspeed increase, or upgust, the nose will pitch up; when going from a headwind to a tailwind, or down gust, the nose will pitch down.

So we're taking off into a headwind and suddenly, right off the ground, we get an airspeed loss, or downgust. Airspeed drops and so does the nose. We should be prepared for this and, as in all flying, hang on to the pitch attitude we want.

✛   Thermals

There's another effect which isn't strictly shear but acts in much the same way. It comes from thermals, the kind glider pilots look for: rising pieces of air.

On a summer day we approach a runway. Perhaps there isn't much wind at all, but before reaching the runway we get a bit of sink and have to pull the nose up and add power. Then we cross the runway and suddenly seem to have excess lift and float for a long distance before the airplane will touch down. What happened?

Chances are that this was an approach to a paved runway surrounded by grass or other vegetation. If we could see air circulation, we'd notice lifting air coming up off the paved runway and then (because when some air goes up, some other air must come down to replace it), we'd probably see air sinking over the grass adjacent to the runway. On our approach we'd fly through the sinking air over the vegetation; then, over the runway, we'd fly into the lifting air that would want to keep us up. The opposite will occur at night, with lift over the warmth-retaining vegetation and sink over the cooling runway. That's why we sometimes go clump in a hard landing at night and wonder why. We flew into sinking air. The night effect, however, isn't as strong as the day.

The important part of all this is that when sink occurs short of a runway, be it due to speed loss from shear or a thermal, we must be ready to use power and keep our airplane flying on the approach path which we have selected, either visual or instrument, and not allow it to land short.

12

VFR—Flying Weather Visually

Some of us fly VFR because we don't know how to fly instruments. There isn't any doubt that a pilot should learn instrument flying and obtain a rating as soon as possible. It makes flying safer, more useful, and more enjoyable.

Knowing instrument flying makes for safer flying because a pilot doesn't get trapped in bad situations. If cornered, you can pour on power, pull up to a safe height, and think out what is best to do. It isn't necessary to poke along between mountains or duck TV towers when ceilings and visibilities are low. Most of the time, an instrument-rated pilot can be on top, in sunshine, enjoying the flight, rather than nervously and dangerously picking along in reduced visibility.

In the early days of the airmail—the days of open cockpits, helmet, and goggles—the pilots didn't fly instruments; they dangerously stayed "contact" and strained eyes and nerves trying to see what was coming up. That was when the Allegheny Mountains were called Hell's Stretch. Pilots plowed into them regularly, leaving broken bodies and airplanes on the rugged slopes. In the first year of the airmail service, as many pilots were lost as the service had.

It pays to be instrument qualified, and every day it becomes more necessary as the airway system grows in complexity. Modern aircraft equipment has become more sophisticated and compatible with weather flying so that instrument flying is an integral part of all flying. To restrict oneself to VFR flight means one will have less and less area to fly in, and someday VFR flight may only be possible in limited zones under strict discipline. Aside from all that, an instrument-qualified pilot

145

is a safer pilot, and it is difficult for me to visualize being in the sky without the ability to fly by instruments.

Night flying, surprisingly, is often instrument flying. Take off at night, down a lighted runway, then, as the runway falls behind, we are suddenly in a black hole with no visual reference. We're on instruments! The record shows that more VFR accidents occur at night than in the daytime. If you're going to fly at night, you'd better know how to fly by instruments.

✈ VFR

But there's still a lot of VFR flying; most flying is VFR, really, and there isn't anything wrong with it if the limitations are known and followed. First, always, is the fact that in flying VFR one must SEE! Low clouds and poor visibilities make this difficult. Rough terrain makes it worse. A safety key to VFR is not to get in a position where the visibility is so bad you cannot see enough to make a normal 180-degree turn and get out. It's worth trying on a clear day, noting points on the ground, to see how much room it takes to turn, then double that distance for a cushion while one reacts, in actual weather, to the idea that the visibility is too low and they'd better turn around.

✈ The Famous 180

Two things are important. One is that the weather will not always have a nice, clear-cut edge that separates VFR from IFR so that a 180 will bring us right back to VFR. While we've been sneaking along VFR, the IFR weather may have been sneaking along right behind us, and when the 180 is accomplished we're still face-to-face with marginal to below VFR. Keep an occasional watch behind to see what's sneaking up. Point two is that limited instrument ability with little instrument instruction and being shown how to do a 180 is not sufficient to get involved with IFR flying. That minimum 180 instruction was for a dire emergency only! With rapidly deteriorating weather, a simple 180 can turn into a full instrument workout and we'd better be able to handle it. If committed to a 180-degree turn, and the visibility reference is poor, concentrate on the instruments: holding altitude, maintaining air speed, keeping a proper bank angle until the return heading is achieved. But remember, without full instrument flying ability, this is very dangerous stuff!

Flying into the picture: IFR yes, VFR no. It's obvious the weather ahead is worsening. This was near Tulsa, Oklahoma, with a summertime warm front and thunderstorms. A pilot should have a good picture of the weather and actual reports before poking around in this.

In VFR flying, it is especially important to be flying toward better weather. VFR flying demands a good knowledge of the weather situation, and a VFR pilot should study weather in advance as much as an IFR pilot. A nice day at the takeoff point doesn't mean it will be nice for the entire route.

The VFR pilot must be especially aware of any possible weather deterioration. If it goes sour, staying "contact" is a must, and this can mean getting too low in reduced visibility—a perfect formula for trouble! The IFR pilot is interested in trends, but deterioration may mean simply shooting a low approach or going to an alternate. The IFR pilot does not get into trouble as deeply or quickly as the VFR pilot when the weather goes sour. So the VFR pilot needs a good look at the synoptic situation: where fronts are and their expected movement, plus the en route and terminal forecasts. And remember our admonition about areas marked MVFR, Marginal VFR. Unless the trend is for improvement, stay out of these areas and, at best, keep a sharp eye on what the weather is doing when flying through MVFR areas.

Basic weather rules for a VFR flight are these: Fly toward improving weather; do not fly toward approaching fronts; especially do not attempt to "beat" a front to your destination. Be extra conscious of destination forecasts when flying toward evening, shorelines, heavy industrial areas, and mountainous regions. Also take special care in spring and fall when the days are warm and the nights cool, particularly if there's a lot of moisture on the ground and in the air near the ground, which we get when lots of rain has fallen, and particularly when there is snow on the ground which has been rained on or subjected to melting during the day. We all know the feel of those clammy, cold nights; they make fog easily.

If there isn't any frontal condition, but cloud decks are forecast because of postfrontal air masses, be alert to the fact that clouds in mountainous areas will tend to be overcast and close to or on the mountaintops. If the air mass is vigorous, there will be showers and low visibilities. The weather will be as difficult as a front.

Flying in mountainous regions requires more visibility than in flat terrain because clouds tend to hang on mountains and blend in, so that mountains and clouds often look alike. The slopes are difficult to see, and one comes up on them fast, even at 100 miles an hour.

A good rule is that if you cannot see the mountaintops, with space between them and the cloud bases, it's a poor time to be flying. There are, however, some conditions where there is excellent visibility, 20 miles or more, and we can fly down valleys with the mountaintops in cloud. But the forecast must be solid gold to stay good! The first snowflake or raindrop and it's time to find a place to land. And always be certain there's an entry and exit to the valley, because you cannot climb over the mountains to get out. Flying into a cul-de-sac—a dead-end valley—is bad, bad news!

When checking weather for flight across mountainous areas, be aware that the METARs—actual reports—may show adequate ceiling and visibility at departure and destination airports but lack reports for areas en route that may be much higher than departure or arrival airports. There might be adequate ceiling and visibility at departure and arrival, but zero-zero in high terrain en route.

☦ A Point to Remember

A VFR point worth thinking about is that the closer one gets to the cloud base, the worse the visibility. As the ceiling squeezes a pilot lower and lower, the tendency is to try and stay as high as possible. In doing so, the airplane crowds the base of the overcast. (This is illegal.) But the base of a cloud isn't clean-cut; mist and shredded hunks of vapor, like a gossamer veil, hang down below the solid part. Pilots can be flying in this area and see the ground straight down, but ahead, because we are looking through a lot of ragged stuff hanging from the cloud base, the visibility is low or nil. Descending a little, sometimes only 50 feet, the visibility may suddenly improve. If it doesn't improve in 50 to 100 feet, then the bottom of the cloud isn't the problem, and it will not pay to go lower; it might actually be dangerous because of getting too close to the ground.

✢ Snow Is Different

This idea of going a little bit lower to improve the visibility doesn't hold true in snow. Often, in snow, the ceiling is quite high. Snow generally forms directly from water vapor without any cloud process, and one doesn't get "in" a cloud. In professional jargon, this process is called *sublimation*. Because there isn't cloud in this kind of snow, there generally isn't any ice either. It takes the supercooled water in a cloud to make ice: no cloud, no ice (except in the case of freezing rain). Therefore, if a pilot is in snow with the ground visible, she or he can pretty well relax as far as ice on wings and propeller is concerned.

Snow can come from nimbostratus, and snow showers from cumulus. These clouds are easy to avoid—generally by flying on top. Nimbostratus, or stratocumulus, are the kind we find over mountains after a cold front. Snow showers in cumulus would be in unstable air, so on top, again, is the place to be, but tops may be high. The clouds that have snow must contain supercooled water, and they get that by the lifting caused by strong instability or in air that's been lifted mechanically by wind pushing it up a mountainside. But generally, in large snow areas there isn't ice. If ice forms, you are either in a lower cloud deck, which you can top, or you are near the frontal surface. More on that later.

The heat of the engine and windshield may cause snow to melt and refreeze as slush, just as on an automobile windshield on a snowy day. Because of this a pilot has to keep an eye out for carburetor ice all the time and for ducts getting plugged up after a long period of flying in snow.

When flying in snow caused by overrunning air ahead of a warm front, in other words a general area of snow, one can see the ground straight down from quite a few thousand feet. Of course, the visibility ahead will be very poor because we're looking through a lot of snow. Because the snow forms aloft and falls to the ground, the restriction in visibility that it causes takes in all the sky; consequently, going lower will not improve visibility. People are sometimes fooled by snow and fly lower and lower because they can see the ground straight down and think they should be able to get "under" the stuff. In fact, they cannot get "under" it and sometimes are hurt trying to. A pure VFR pilot had better stay out of snow areas because the visibility will be poor regardless of the ceiling. Once snow starts, even light snow, the visibility will become poor in a short time.

A VFR pilot prowling along under leaden skies will often have good visibility, but when snow starts and the first white streaks go by, it's time, right then, to think about going elsewhere or finding a place to land. The first flakes of snow are almost always a loud and clear signal that things are going to get worse. It's no time to keep poking ahead hoping for better.

Sometimes, ahead of warm fronts, snow begins to fall very quickly in a fairly wide area, 100 miles wide or more, so that one must consider the unpleasant possibility that a 180-degree turn will not get us out of the condition because the snow may start, at the same time, all around us and for many miles behind. All this relates back to the point that we should study weather in advance and know, in our mind, what could happen and which way to duck if it does. When precipitation begins, especially snow, a pilot shouldn't sit fat, dumb, and happy until things get worse, it's time to do something before that; either try to turn around or else go somewhere and land. Which is a good argument for keeping position: knowing where we are and where the nearest airport is.

✝ Keep Calm

When we say go somewhere and land, we don't mean to do so in a panicky fashion. The worst thing we can do, in any situation, is get frantic, dash for Mother Earth, and land in the first field that comes along. If possible we ought to stick to airports, and ones that fit the airplane. Of course, this isn't always possible, and more on that later.

✝ More Snow

Snow, just to confuse things, can also be found falling out of clouds. This condition is generally found in air-mass weather, such as a stratocumulus deck behind a cold front. Snow falls from the clouds, and the clouds have ice in them. In mountains, the higher ridges are in the clouds, and these are conditions a VFR pilot should avoid.

The mountains of Pennsylvania, after a cold front passage in winter, are generally cloud-covered and getting lots of snow, and the visibility is poor to zero. It's not a frontal condition, but nevertheless it's a tough one for VFR. It's not a place to try to fly VFR underneath. The only way is on top, and that will require an instrument climb with ice in the clouds—tops eight to twelve thousand!

Low stratus clouds, giving poor ceilings, increase as one gets closer to a front. In these clouds there is snow and ice; the snow, however, will be falling through the clouds from someplace above them. A pilot will be in snow, but this time the ground will only be visible occasionally through breaks. There will be ice in the stratus clouds. At best this isn't a good place for VFR flying. We'll talk more about it later when discussing instrument flying in weather. We can say, in passing, that a pilot can pull up and get on top of these ice-producing clouds; flight will be between layers but mostly on instruments, because the snow will cut visibility practically to zero. It will be a white-gray world without reference—for instrument pilots only.

✝ TV Towers

VFR pilots sneaking along in flat country have to be wary of TV towers that stick up dangerously high and are difficult to see in conditions of reduced visibility. The TV people also have a way of putting towers on mountaintops, and it can be quite a shock to sneak over a mountain with clouds crowding the top, only to be suddenly confronted by a massive TV tower and its supporting wires.

✝ Maps Are Important

TV towers are marked on maps and can be spotted if the VFR pilot knows where the airplane is, which brings out the important point that VFR pilots must be sharp navigating by map, and use one! Navigation is most difficult down low, because you don't see as far and there aren't as many clues available to help interpret a map; at low altitude the area your vision takes in is smaller, even with good visibility, and, of course, in bad weather the visual range is even worse.

Sneaking along VFR requires moment-by-moment knowledge of where one is; a pilot who doesn't always know position should raise the ceiling and visibility requirements to a higher value, say 3,000 feet and 10 miles as a starter.

Constantly knowing position also means knowing where the nearest airport is to run for if things get tough. It may be one recently passed, off to the side, or not far ahead, but we ought to know where it is. The GPS (Global Positioning System) and LORAN most people have on board now, have the handy ability, by simple keystrokes, to give direction and distance to the nearest airport. This is great stuff, but I emphasize, should not take the place of a pilot constantly knowing where the airplane is and how high the surrounding terrain is, and ahead, too!

✛ Not Only Airports

If we're in a jam, stuck with IFR facing us and no place to go, there may be relief right under us. It isn't always necessary to land on an airport, and the "gotta get to an airport" fixation may work us into a deeper jam than picking a field and landing in it.

This, of course, demands care and the realization we may bend a nosewheel. But glider pilots land in fields almost routinely, and how to do it is part of their training. In glider talk it's called "off-field landing," meaning off airport. And knowing how to do it is another reason for power pilots to get some glider training and experience.

First, take it easy and take your time. Don't rush! Look the field over for rocks, ditches, large wet areas, and severe undulations. Check its slope, both the long way and side to side. Landing down a slope is tricky, and landing with a strong slope to the side is also. Check carefully for obstructions, especially wires that may be concealed by trees. Then when it's decided to land, do a proper and normal pattern: downwind, base, and final, all the while looking the field over for any further obstructions and rough surface that may become apparent as you get lower.

But landing away from an airport in a field may be far safer than trying to work along in bad visibility and ceiling with the ground getting closer and the visibility ahead poorer. But stay cool!

There's a great tendency for people to tune omnis or ADFs and fly along on the signals, knowing they are headed for the right place but never really knowing where they are! A radio facility chart doesn't take the place of a map! Maps, too, tell us important information about terrain heights, mountains, and valleys, and a host of other things that a VFR pilot should know in detail.

To have useful terrain height information, a pilot must keep the altimeter set to the barometric pressure of the nearest airport so the altimeter will tell with accuracy where we are in relation to the terrain.

If things get really tough and we are prowling down low, perhaps following a railroad, we have to keep in mind high-tension wires that may cross the valley that we and the railroad are following. Down that low, heaven forbid, we also find radio masts that aren't on maps and other things that stick up in our sky. Crowding a mountainside too closely could possibly cause one to clip the wires of a ski lift that may be difficult to see, summer and winter. A number of glider accidents have occurred in the Alps in exactly this manner. This kind of flying is desperate, not to mention illegal, and no one should be doing it; I mention it only in case one gets oneself in such a fix and, illegal or not, wants out in one piece.

✢ Where Is the Wind?

Flying VFR, a pilot must be constantly aware of the wind, where it's coming from, and how fast. In mountains it's helpful to realize that there will be more cloud on the upwind side of a ridge, and it doesn't have to be a steep ridge. Even a gently sloping hill will give enough lift to the air to cause clouds.

On the downwind side there will be downdrafts, and if the wind is strong, they'll also be strong and turbulent. One needs to be flying in clear air, with visibility, when one wrestles downdrafts, so fly a fair distance from the mountain on the downwind side where it will be clear or the ceiling higher. The airplane's altitude should be above that of the range, if clouds will allow. The worst downdraft spillage on the downwind side of a ridge will occur at or below the level of the mountains, perhaps a little higher than the ridge, but not much.

There are areas where winds converge. This can cause turbulence and, sometimes, enough lifting to make clouds. If the wind is flowing down a valley and there's a ridge in the valley, the wind will split as it goes around this ridge. On the downwind side, the wind comes back together again, pushing against itself, creating an area of convergence and vortex.

We have to consider that cuts in mountains, ridges oriented in different directions, peaks, and valleys all cause moving air to tumble and eddy in very irregular ways that are difficult to visualize. Wind coming down one valley and flowing into another will give turbulence and convergence. Caution is called for in mountainous areas, especially in areas where the mountains are arranged in hodgepodge fashion and not neatly lined up. There can be areas of cloud in unexpected locations because of mountain irregularities.

✢ Near Cities

If one is approaching an industrial area, town or city, a detour around on the upwind side will give better visibility. This rule will often apply to bodies of water, too, where a light wind can give more fog and lower visibility on the downwind side.

✢ Summertime

While summer is a nice time to fly, it sometimes brings serious visibility problems to the VFR pilot. The sky may be cloudless, but the summer haze makes a cross-country flight by map reading and looking a difficult chore. This is an especially good time to be on the upwind side of cities, because the polluted air mixes with the haze and makes the visibility much worse.

An inversion causes all this, where the temperature aloft gets suddenly warmer. The hazy air can rise no higher than this altitude where the temperature increases.

Climbing, we come out of the mass of glop and find ourselves in clean, clear air with miles and miles of visibility. Looking down, however, we see almost nothing, and navigating is difficult. An occasional river glints through the smaze, the white ribbon of a highway, a piece of a mountain. It's all difficult to put together in the jigsaw puzzle of map reading. If we are omni, ADF, Loran, or GPS equipped, we can use this equipment for navigation and by looking at the occasional clues below know where we are, where that nearest airport is, and how we are doing. Without navigation systems help, it's a job of superconcentration, and often it's worthwhile to swallow pride and fly a slightly longer course to follow a visible highway.

✢ Thunderstorms and VFR

Summer brings thunderstorms, of course, and the reduced visibility in haze makes them difficult to see. On top of the haze level, the big cauliflower clouds can be seen clearly and be avoided. Down low, in

The effect of air pollution on visibility. This smoke extends many miles downwind, while upwind it is clear—showing why it is obviously best to fly upwind of cities and industrial areas. (NOAA PHOTO)

the smaze, one doesn't know if the bad visibility is smaze or darkness from a closeby thunderstorm. Flying high, above the haze, ducking thunderstorms, a pilot needs to watch below, too, because clouds may sneak in under if we're close to a storm.

There are ground-based radars that FSSs and the NWS have available that show thunderstorms and their location. A call can get you a pretty good picture that's quite current. Ask plainly for what their radar is showing. Then use that information for *avoiding* the areas, not for squiggling through them! ATC may not always be able to provide the information because of traffic demands, so don't take off planning to fly through a thunderstorm area by getting help from the ground.

Thunderstorms, at best, are things to keep away from. This is especially true for the VFR pilot. A good look at the weather before flight will show if thunderstorms are going to be of the air-mass types or ones caused by frontal activity, and therefore, how to plan around them. Air-mass storms are scattered enough to tour around, staying in the clear. If they are frontal storms, the VFR pilot's place is on the ground!

Ducking around thunderstorms means staying out of them, and it means, too, giving them a wide berth. Turbulence outside a storm can be as rough as that inside. There's also danger of hail falling from the high, overhanging anvil portion of the cloud into clear air. A general rule is not to fly over lower clouds or under higher clouds in the immediate vicinity of thunderstorms.

While weaving around, it's best to keep working upwind of the storm because most action is on the downwind portion. Also, an eventual return to course will be shorter with faster groundspeed. Compass headings should be noted: Knowing the time on various headings is a help in guesstimating how far off course one may be going and which way.

An important point about headings is that if one seems forced to keep working in one direction, say southwest, and is unable to work back northwest because of storms, it's obvious that one is making a big end run. This means that the storms are pretty well lined up and there isn't any clear path on the desired course. It may be a front, a prefrontal line squall developing, or a bunch of air-mass thunderstorms that have lined up in a pseudofront. It's time to land or turn around and go back.

A pilot must consider that the storms being circumnavigated may be right over the destination. If so, it may be necessary to go elsewhere or else fly around out in the clear until the storm drifts away. This latter means there must be enough fuel on board and daylight remaining.

Flying VFR at night around thunderstorms isn't a recommended procedure; the difficulty of seeing where clouds are is great, and in early evening the storms are probably at their worst and clouds more extensive. One simply should not be out there unless instrument qualified; it's an instrument environment.

A point to remember about air-mass thunderstorms is that they occur mostly in the afternoon. If a pilot gets an early-morning start, it's probably possible to reach the destination before afternoon storms have built up and covered a wide area difficult to get through.

There is much more to be said later about thunderstorms, but right here, while talking about VFR, these points are important, and the most important of them is to stay in the clear and always have a wide avenue handy for hasty retreat.

✝ VFR on Top

Flying VFR doesn't mean one has to stay under clouds. If there are scattered clouds with a top that's not too high, it's better to be on top in clear air than down in smazy air worrying along in reduced visibility. It's also more pleasant because the air is smooth and cool.

You must be certain, however, that those scattered clouds stay scattered and don't become broken or overcast. A noninstrument-qualified pilot must be able to come down without going through clouds.

A good study of weather before takeoff can assure the pilot that the flight isn't toward frontal conditions or into mountainous areas, where clouds tend to be broken and at times overcast instead of scattered with higher tops. But if the clouds are scattered, there's no sense in sitting down low working and worrying unnecessarily.

✝ Use the Radio VFR

Most airplanes are radio-equipped, even though they may have only simple equipment. The VFR pilot can use this equipment, however, as an aid to navigation, for getting weather reports, and for following a VFR flight plan. This is good training for the future when working to obtain an instrument rating.

There are two basic parts to instrument capability. One is knowing how to fly by instruments, to keep the airplane under control without outside visual reference. The second part is navigation by radio, handling radio communications for ATC purposes, and obtaining weather information.

The pilot who cannot fly instruments can nevertheless learn and develop proficiency in navigation, ATC, communications, and weather gathering while flying VFR. By listening to scheduled weather broadcasts, we learn when they are broadcast, where from, and on what frequencies. We can develop good habit patterns and useful knowledge. We can navigate using radio, following a track, crossing checkpoints, and checking speed, remembering, of course, to keep tabs with a map, too.

We can file VFR flight plans and keep in touch with Flight Service Stations en route, obtaining the altimeter setting of the nearest airport so height is correct. Give an estimate for destination, and requesting weather information is all "free" training.

If VFR pilots do all these things, they will develop a facility that will be useful in instrument flying. Practicing and learning VFR will make passing the instrument test easier and, most important, will mean that when starting to use a newly acquired instrument ticket, things will be familiar, useful, and safer, sooner. There's a lot of value, present and future, in using many IFR procedures while flying VFR, those within legal bounds, of course.

✠   Without Radio

Pilots without radio can fly VFR and certain weather, too. I flew my first seven years without radio of any kind, and this included flights coast to coast, to Mexico, and to Cuba.

Without the availability of en route information, the preflight study must be thorough. In-flight observation of clouds and their type, visibility, wind, and any other phenomena that may signal changing conditions becomes a constant and crafty task. Occasionally it may be necessary to land and check just what the heck is going on.

But the game is interesting, and a pilot may learn much so that someday, when instruments and radio are available, that pilot will be a better weather pilot.

Two points not to be forgotten when flying VFR are the legal minimums and where one can and cannot go VFR. Like all minimums, they should not be our only guide; there are times when it isn't comfortable flying even though legal minimums exist. Comfortable means just that. When you are a little tense, wishing things were better, your psyche may be telling you it's time to quit, VFR minimums or not.

While flying VFR there is a strong responsibility to look for other aircraft. Often VFR flying is putting us where many IFR aircraft are flying: on top, between layers sometimes, in traffic areas where instrument approaches may be coming out of a legal VFR ceiling. Some of the IFR pilots, unfortunately, have a mistaken idea that because they are in the system, on a flight plan, they are protected from all aircraft. Well, they aren't protected from VFR traffic in many places, and this puts a burden on the VFR pilot to keep alert, looking for that other aircraft that could spoil your day!

So, now on to the pilot with an instrument rating, a piece of paper that doesn't necessarily mean one is a weather pilot because, at first, it's only a learner's permit.

13

About Keeping Proficient Flying Instruments

We may have an instrument rating, but unless we stay proficient, the rating isn't really any good. How do we keep sharp? Airlines, of course, check their pilots extensively in simulators. They also are observed on actual flights by check pilots to see how they do the entire job and manage a crew.

If an individual wants this kind of checking, there are places to go and pay for it. Both check-flights and refresher courses are an excellent idea. But aside from going to professionals, there are ways we can stay at top level and test our own abilities.

✛ Practice

First of all is the way we fly every day. What we should do is fly precisely and practice in good weather or bad. If we fly as though an FAA inspector was looking over our shoulder, we'll fight off complacency and sloppiness. Surprisingly, trying to fly the best possible way will finally become our normal flying character, making us better pilots all the time.

This means holding altitudes and headings as closely as possible, flying precise airspeeds in climb and descent, and making corrections smoothly and exactly. If we do these things, they become a habit.

An important part of this is to use the ILS every time there's one on the runway we're landing on, good weather or bad. The more practice, the better.

It's also interesting to shoot ILSs on a VFR basis occasionally to see where the ILS takes you. One learns about bends in both localizer and glideslope and gets a chance to see what various altitudes and terrain clearances look like along the ILS path. We can learn how far "off course" it's possible to be and still make a turn to get on the runway or, even more important, how close one must be to the localizer at minimum altitude in order to have enough room to turn and get on the runway.

Different ILSs act in different ways. They have their own characteristic bends. Also glideslopes are not good all the way to the ground. Their signals become useless at some altitude near 100 feet. I've found that this altitude varies with individual glideslopes; some are good down to 80 feet, while others "come apart" at 150 feet.

By practicing approaches, we can see where the glideslopes at runways we frequently use lose their accuracy. All this can be handy knowledge on some stormy night when we really have to shoot a tough one. Into the airports I use, I try to know each ILS and its peculiarities.

Proficiency is a matter of practice, and that is a matter of doing. Almost the only pilots I've ever seen who have trouble on instrument checks are the ones who get in the habit of doing only the necessary flying and no more. If it's VFR, they just go in and land. They don't make use of flying's opportunities. I've seen fellows have trouble and then, when awakened by the jolt of a down check, get to work and come back to top-level proficiency. Generally, one jolt lasts a lifetime. It not only raises the ugly idea of being without work, but it's a real injury to one's pride.

✝ Self-checking

We are also interested in how good we are, what sort of situation we can handle. Well, in the absence of a check pilot, let's be our own check pilot.

The first thing we need is someone to ride along who will look out for traffic and who can recover if the airplane gets out of control. Two pilot friends can ride for each other and learn a lot in the process.

Once in the air and at a good altitude, 4,000 or above, put up the hood and cover the horizon and directional gyro. Then do these exercises on the primary flight group: needle-ball, airspeed, and vertical speed.

The exercises should be done without any appreciable altitude loss, overshooting or undershooting altitudes when descent or climb is called for, or when there are great airspeed fluctuations. Any indication that you don't have the airplane under control, that you are unable to keep it from stalling or exceeding its high-speed limits, means instruction and practice are needed before you do any instrument flying.

So here are the test maneuvers, under the hood, primary flight group:

1. Straight and level for five minutes.

2. Climb at 300 feet per minute—airspeed 50 percent above stall. Climb like this for 1,000 feet. Level off at an exact altitude. Then descend at 500 feet per minute, same airspeed, for 1,000 feet and level off where you started the original climb.

3. Do a straight-ahead stall and recovery. A real stall! Not an approach to stall, but one that makes it shudder and shake and want to duck a wing.

4. One-needle-width turn to right for 360 degrees. Maintain altitude within 50 feet. Roll into one-needle-width turn to left for 360 degrees. Hold altitude.

5. Stall from 30-degree banked turn.

6. Establish three-needle-width turn to left for 360 degrees. Keep turning and start descent, allowing airspeed to build up to 30 percent above cruise IAS. (Don't exceed any limits.) Descend 1,000 feet. Stop descent, level off, and come out of turn.

7. Following above, immediately roll into three-needle-width turn to right. Do one 360-degree turn and then climb in the turn at 70 percent above stall IAS for 1,000 feet. Level off and stop turn.

If you can do all these things neatly and with precision and be master of the situation all the time, you are a pretty good instrument pilot.

✝ With Full Instruments

Now uncover the DG and horizon, fly with full panel, and do the following exercises. They are to show the precision demanded when flying in a traffic area and executing an approach and pullup. They should be done with great exactness: airspeeds within a few knots, altitudes exact and not varying over 50 feet, headings hit on roll-out within 3 degrees.

1. Straight and level. Hold exact heading and IAS.

2. Reduce speed to 50 percent above stall. No altitude gain or loss.

3. Reduce speed to 40 percent above stall. Turn 45 degrees to right. No altitude change and hit new heading within 3 degrees on roll-out.

4. Maintain above speed and descend at 300 feet per minute. Turn left 170 degrees. Lower gear.

5. Level off, lower flaps to landing configuration, reduce to approach speed. Hold this for two minutes.

6. In above configuration, descend at 500 feet per minute to a preselected altitude 1,000 feet below.

7. At that altitude do a pull out. (Tune omni to new station and set in a different radial.) Turn left 90 degrees as you do. Establish climb speed and clean up gear and flaps.Climb 1,000 feet and level off.

These exercise maneuvers are to check your sharpness. They aren't an instrument pilot's final exam. They are simply something to go back to now and then for an appraisal. The degree of precision obtained says how good you are.

The maneuvers on the primary group are really designed to make certain you can keep out of, or recover from, a spiral condition with increasing airspeed and recover from a stall.

One can vary these maneuvers and design more for oneself. The basic thing in the full-panel maneuvers is to try and load oneself up with all the things that might be required and see if they can be done with the precision demanded by the ATC system and, of course, safety. This means one needs an ability to change speeds while climbing or descending, turn to headings, lower flaps and gear, adjust rates of climb or descent, and hit altitudes and headings exactly. There's also a requirement to remember these altitudes and headings and keep them after leveling off.

One could do all these things and toss in an extra job of consulting charts and worrying about carburetor heat, as well as talking on the radio. It can be quite a load, and we'd better face the fact and be prepared to handle it because that's the kind of load one can encounter in the system. It is a handful for a neophyte instrument pilot alone and without autopilot.

We should check ourselves periodically—every few months. An honest self-appraisal is often more severe than a check pilot's. This isn't the place to kid ourselves; if we're not up to snuff, we should practice more or go back for help from a professional instructor, or both.

14

Thunderstorms and Flying Them

Thunderstorms are thunderstorms, and while they may come from different beginnings, they all turn into the same thing. Once a pilot is in one, he or she wants to know how to fly through it and come out safely. Even more important, however, is keeping out of it in the first place; and no matter how old or bold a pilot may be, the primary thoughts about thunderstorms are concerned with staying out of them! To help stay out, or fly through if we must, we should know what thunderstorms really are and the different backgrounds they come from.

✛ What Are They?

Simply a cumulonimbus, CB, thunderstorm, or T, whatever one calls it, it is a concentrated mass of very unstable air in violent motion up, down, and sideways. It has strong, gusty winds that generally extend to the ground and make landing or takeoff difficult or impossible. It has thick clouds, heavy rain, and sometimes hail. Electrical discharges occur frequently. Some Ts have tornadoes associated with them, but most don't. Tornadoes, however, never occur without thunderstorms. Tornadoes are associated with thunderstorms, but they do not always come out of the big, bunched-up cumulonimbus clouds with the highest tops. They often come from much smaller clouds that hang back in a line from the main storm center. So it's wise not to duck under or fly through clouds close to a thunderstorm just because the clouds look smaller and have lower tops.

To make a thunderstorm, we need conditionally unstable air. What's that and who cares? Well, it's air that is stable as long as it doesn't condense...that's the condition. When it does condense, the release of heat in the condensation process makes the air warmer and it wants to go upward. Then more moisture condenses, more heat is released, and it goes up more. It's almost like perpetual motion and the energy released is tremendous.

The normal places we go for weather information cannot tell us if an air mass is conditionally unstable or not. That is learned from upper-air soundings taken in many places over the country and analyzed by a NWS computer. But if a weather forecast calls for thunderstorms, it means the air is conditionally unstable or will be, and that's all we need.

✛ What Is Tough about a Thunderstorm?

What really bothers us most in a thunderstorm is the turbulence. Lightning discharges are a minor danger and few if any airplanes have been knocked down by lightning. In recent years a question has come up of flame going up the fuel vent and then igniting a near-empty tank loaded with explosive vapors, and there is strong evidence that this may have happened. Redesign of the vent system, plus a method of discharging inert gases into the offending vent system when flame begins to move, shows promise of correcting this situation. But as for lightning hitting a plane and knocking it down, the evidence says it's doubtful. I spent a number of years looking for and flying through thunderstorms in research work and saw a lot of lightning. I experienced numerous electrical discharges where the aircraft was one end of the lightning cycle—four in one day—and the worst damage was holes about the size of a half dollar burnt on the trailing edge of control surfaces. Bigger holes have occurred, but they generally aren't caused by the lightning. When I had a discharge in a Boeing 707 out of London, England, a three-foot-long piece of the black radar nose was torn out. The lightning actually made a much smaller hole, but the high-speed air from our forward motion got inside the nose and then tore out the bigger piece. There wasn't any special problem, but I dumped fuel and returned to have it fixed before crossing the Atlantic.

✛ Tornadoes

A tornado, of course, would be catastrophic if an airplane flew into one. The winds are unbelievably violent. But tornadoes can generally

*Damage to a Boeing 707 from electrical discharge, often called a
lightning strike. The point of discharge is small and near where
fuselage and radome meet. Large, torn pieces were probably
ripped out by airstream. None of this affected the airplane's
flying ability.*

be seen if one is outside cloud, flying VFR. The big hazard is sneaking
under thunderstorms at night or down low in and out of cloud. There
is always the chance that while you are sneaking under clouds near a
thunderstorm, a tornado may form, perhaps where you don't see it. I
was lucky one night, in a DC3, flying from Chicago to Indianapolis
while trying to squirm my way through a line of thunderstorms. I was
flying low, as we often did in preradar days, trying to stay "contact."
Ahead I saw a strong glow which turned out to be a large field on fire,
probably started by lightning. But the shock was to see this fire lighting the sky enough to show a tornado with its scary funnel right on
my course! I went around it, somewhat shaken, because what if that
handy fire hadn't lit up the sky! It's always useful to have a little luck
around.

It does seem possible that an altitude above the base of the thunderstorms, perhaps 8,000 feet or more, may be above the tornado. It
will be a ride to remember, but the violently destructive part of the
tornado may be missed. But best, by far, is to avoid the entire area.

A friend of mine passed over the Pittsburgh, Pennsylvania, radio-range station in a DC3 many years ago at 8,000 feet while a tornado was going by on the ground at almost exactly the same place and time. He didn't have a bad ride and flew on to Washington, D.C. There have been numerous cases of this sort, so there's more than just theoretical evidence that the violence of a tornado is in the low levels only. Part of the reason for my friend's good ride was that he probably missed flying through a cell.

One criterion, developed by Dan Sowa of Northwest Orient Airlines, is that if the reported radar top of the thunderstorm pushes up above the tropopause by 10,000 feet, the storm has tornado potential. One can get thunderstorm and tropopause tops from reports via NWS, FSS, and Flight Watch. But I do not feel that the reported tops of all thunderstorms are accurate. I have always felt that knowing the tropopause height is a better indication of how high the thunderstorm tops will be: the height of the tropopause plus a few thousand where the storm's energy pushes it up into the stratosphere. And few, if any, thunderstorm tops are less than the height of the tropopause; it's the only inversion to shut them off.

� Hail

Hail is always a problem, but compared to the numbers of airplanes flying in and around thunderstorms, not many are hit by hail, and there isn't any record of an aircraft actually being knocked down by it, although a few have been pretty well beaten up. There is evidence that some hail can damage a jet engine so it loses its effective power. Apparently, one catastrophic forced landing resulted from this.

Most hail comes out of the big overhang cloud that is downwind of a thunderstorm; if it's visible, this is the place to stay away from, even if it looks clear and pleasant there. You might get right under it when the hail dumps. Thunderstorms demand a wide, respectful berth!

Heavy rain can cause engine problems in the form of carburetor icing, and all precautions must be taken. Sometimes the rain seems so heavy, in a thunderstorm, that the engine should simply drown out. There is still some question about jet engines in heavy rain, so it is important to follow recommended procedures meticulously—such as ignition on, when called for.

What hail can do to an airplane. But it flew! (NOAA PHOTO)

✝ The Bad Part

The big problem, as we said, is turbulence sufficiently severe to make keeping control of the airplane difficult. The turbulence may be severe enough to tear an airplane apart, but it is most likely that structural failure follows loss of control and subsequent high airspeeds that may reduce the airplane's structural integrity.

Airplanes are pretty tough and, if flown properly, will probably traverse a thunderstorm safely. This isn't meant, in any way, to suggest that pilots fly through thunderstorms on purpose. It is and always will be best to stay out of them. There is some evidence that if certain severe thunderstorms are encountered at the point of their most potent growth, it may not be possible to maintain control or keep the airplane in one piece.

Landing with thunderstorms near or on the airport is a hazardous operation. Most landing shear accidents have occurred with thunderstorms nearby, and some not so close, to the airport. Near the ground, as on final approach, is the most difficult area to handle thunderstorms. There's no altitude buffer in which to recover, as there is up high where altitude can safely be lost during the rough ride. Close to the ground, an uncontrolled altitude loss is disaster.

In the past we've always known, through observation and experience, that winds close to a thunderstorm are squirrely, gusty, unpredictable in direction, pushy, jerky, violent, and sometimes cover a bigger area than imagined. We learned it wasn't smart to get too close, especially in landing—and not to take off into such situations.

Today research has analyzed and put names on all this: microburst, macroburst, gust front, downburst, and so on, but this hasn't changed the action; it's all out there, just as vicious as ever. Unfortunately there has been little advance in finding these conditions, locating them precisely, quickly, and then getting the information to the pilot. Quickly means a matter of seconds in some cases, so that no matter what system may come in the future, Doppler radar for instance, it's of no use unless it gets to the cockpit fast! The discovery and location should be done and announced right in the cockpit, not via some intermediary such as the tower. As yet there's no instrument that will do this.

Instrument systems are available to help fly through these conditions, but they are not a panacea for landing in or near thunderstorms, and there are some storms and conditions the airplane, instruments or not, cannot handle!

The classic wind shear profile is first a headwind, which boosts airspeed and tends to make the pilot reduce power, followed by a sudden loss of airspeed, which then requires scrambling to get lots of power back on, plus an increase in angle of attack—pulling back on the wheel or stick—to combat the severe sinking. The trick is to pull back to get maximum lift, but not so far as to stall the airplane.

The cockpit instrument systems help the pilot fly through the shear with precision, but without instrumentation the pilot is just searching for something in the dark. But don't get the idea these instruments are to find and announce the condition in advance. When they sound off, you're already there!

But there's another important aspect of the nasty winds, and that's the vortex part of a strong downburst, microburst, macro, or whatever you want to call it, which often is encountered after the classic shear. This vortex area is where the wind currents are turning and tumbling without any organized pattern or sequence and, to use it again, can best be described as squirrely. This in turn means the airplane can be battered up, down, and sideways in a matter of seconds without any comprehensible order, and there aren't any instruments for that! The pilot just pushes, pulls, and twists, trying to handle what's happening so fast it's about impossible to analyze and fly it.

When approaching to land, it also isn't likely that Mother Nature has politely set the microburst, shear, or what have you exactly centered on course. So we may hit the side, corner, or who knows what part of the violent winds. This will, of course, confuse the action pilots think they are taking.

What all this says is that in some cases we are at the mercy of the tempest. Obviously it is best to avoid it, and I repeat: do not land or take off with a thunderstorm on or near the airport. This brings up the negative side of technology. In this case there's a mistaken idea, conscious or subconscious, that modern radar can show thunderstorm boundaries well enough so that we can fly close to them and ATC can vector us close also. Not true! Distance required from a thunderstorm hasn't changed, and radar—still a gadget that only shows rain—has not improved enough to show the sharp demarcation between smooth and turbulent air. In addition, both pilots and controllers generally have not grasped the fact that a thunderstorm can develop, become vicious, and do nasty things in minutes! So that shaving the edge too close is not only basically hazardous, but there is additional danger in that the air can change from smooth to impossibly rough in moments!

It isn't always easy to know if a thunderstorm is close or on the airport. The NWS-FAA recognition and then warning to the pilot is a slow, cumbersome, and bureaucratic process, and the word may not get to you promptly. Don't count on LLWAS (Low Level Wind Shear Alert System). There have been numerous times when it didn't work for various reasons, such as the wind sensors located off the airport while the shear was on the airport. Also be skeptical of ATIS (Automatic Terminal Information Service) weather information; it can be very old compared to the urgency of rapidly changing thunderstorm activity. It is up to the pilot to be cautious and vigilant, using any means to get the news regardless of outside help. How? Of course, know the general weather situation and use your own radar and eyes to watch carefully for fast development.

Fast development means big CUs or CBs becoming thunderstorms within minutes. If an airport weather report says "towering CUs and CBs," be careful to check that they aren't turning into thunderstorms— especially in the afternoon or with approaching fronts. This can happen so fast that the airplane landing ahead of you may fly through an innocuous rain shower on approach, but when you come in, minutes later, the storm may have fully developed and dumped its heavy rain and the downbursts, micros, and so forth with it. This has occurred just as told here! If there are towering CUs and CBs with rain showers in the area, be extra suspicious and alert for thunderstorms' sudden development. Incidentally, ASOS fails here in that the automatic airport weather reports do not tell you of towering CUs or building CBs.

You'll have to find actual person-observed weather for that—and your own eyes.

This situation is, again, one in which the pilot had better remember that the ground isn't going to be dependable for getting the weather to the cockpit fast enough; it's up to the pilot to be on top of it!

Incidentally, the first time downbursts, as a result of thunderstorm rain, were mentioned was in the July 1945 issue of the *Journal of Aeronautical Meteorology*, by C. E. Buell!

✝ Their Life Cycle

Thunderstorms, like everything else, are born, live, and die. They do it faster than one might think, but as one dies, another forms.

You do not see all this when approaching an area of thunderstorms. Large clouds pile up in great masses, and unless there's just one lone thunderhead of an air-mass or orographic type, there will be high, intermediate, and low clouds as well as the cumulus.

All these clouds aren't wild and rough; most of them are not and contain only light to moderate turbulence. But buried in those clouds are thunderstorm cells, anywhere from a few thousand feet to six miles or so across. In these cells things are rough and wild. They are the heart of a thunderstorm.

The storm may have started as a snappy, jolty little cumulus that formed and started to grow. It continued to grow and finally became so high that it passed the freezing level; there it stopped being a simple cumulus and became a thunderstorm. This is the cumulus stage, and during it the cloud is almost all updrafts, some going up 3,000 feet a minute; there isn't any anvil or rain.

When I was doing weather research, we measured electrical fields and flew a lot of CU and thunderstorms doing it. One early afternoon, over the Pocono Mountains of Pennsylvania, I flew into a large, growing CU that was going through about 14,000 feet but didn't show any signs of being a thunderstorm. It turned out to be one of the roughest rides we ever experienced, however, and the electrical field strengths were tremendous, which says it doesn't have to be a full-blown thunderstorm. The building and early stages are the roughest.

The way to study a growing CU is by careful scrutiny of the very top part of the cloud, the shreds and pieces. If one can see a spilling, growing action, and if it's "busy," there's lots of action inside and it's growing. Any indication of a small, flat, thin cloud around the edges, near the top, is strong indication that this CU is becoming a thunderstorm.

Three stages of a thunderstorm (from left to right):

1. The early or building stage. Lots of updrafts and rough.

2. The mature stage. Heavy rain, possibly hail; up- and downdrafts; chopped-up, rough air; and strong surface gusts ahead of the storm.

3. The drying stage. Mostly downdrafts. Rain diminishing and not very rough. But it still looks that way. (NOAA PHOTO)

The mature stage follows the cumulus. It begins when rain commences. The cloud develops downdrafts as well as updrafts, along with heavy rain. It's still very rough because of the added conflict of air moving up and down.

After the big anvil has formed off the top, the thunderstorm enters the dying stage, the wild currents decrease, and it calms down, although it still looks pretty bad.

That's the way thunderstorms come and go, but from a cockpit we want to know more practical things: Where are the tops, the bases, how far through, what's in there, and very important, what kicked them off in the first place? Because if we know that, we have a start in knowing how to combat the particular storm we're faced with.

There are three basic ways thunderstorms are created:

1. By heating.
2. By a front.
3. By orographic influence such as air moving up a mountainside or sloping terrain.

All of these can unite and produce thunderstorms faster and more violently than usual.

When a thunderstorm occurs because of air-mass moisture and instability, the air mass will be heated faster and rise to its condensation level quicker on the sunny side of a hill, where glider pilots would look for thermals; a flow of wind up the mountain will also add impetus to the process.

A front shoving air aloft on its own will do it faster and with more umpf if that front is climbing up sloping terrain, and even more so if it's doing all this during the hot part of the day. So hills and mountains make thunderstorms tougher, as they do for other weather, and the rule of extra care in mountains applies again.

Thunderstorms are graded by the NWS with a system of numbers called VIP (Video Integrator & Processor). The numbers go from VIP 1 to 6. They are categorized as:

- VIP 1 & 2. Weak to moderate, turbulence possible, with lightning.
- VIP 3. Strong, severe turbulence possible, lightning.
- VIP 4. Very strong, severe turbulence likely, lightning.
- VIP 5. Intense, severe turbulence likely, organized wind gusts, hail likely.
- VIP 6. Extreme, severe turbulence, large hail, lightning, extensive wind gusts, and turbulence.

But they all are potential trouble. A VIP 2 reads, "Light to moderate turbulence is possible with lightning." Well, if there's lightning it's a thunderstorm and if it's a thunderstorm the turbulence is dangerous. An additional very important point is that a VIP 2 can change to a VIP 5 in minutes. In one case I know the change only took eight minutes! So don't be mesmerized into thinking a low VIP level is not too bad. They are all thunderstorms; some are just worse than others.

Many believe, and I certainly agree, that FAA and NWS dissemination of VIP information is misleading and creates rather than cures hazards. The information should be used with caution.

I have also strongly requested of the FAA that no VIP level be given unless a trend is given with it, for example, "VIP 2, developing to VIP 5" or whatever. The report might even show the trend as being better, down from a higher VIP level to a lower one. But the trend is necessary. Too frequently a pilot reads thunderstorms in an area as VIP 2 and feels that it's not too bad to wrestle with, little realizing that within minutes that VIP 2 can become VIP 5!

Gradually, present National Weather Service (NWS) radars will be replaced by WSR-88D NEXRAD, radars which will result in better coverage across the country. NEXRAD (Next Radar) will provide 16 levels of precipitation, but they'll be reduced to the 6 VIP levels we're accustomed to. In the future they may be called DVIP for Digital Video Integrator Processor, so be aware but don't count on it because things change faster than we can write and print books. Also, the NEXRAD radar will not cover all the sky is some mountainous areas. Reading up and studying what's new is the only answer.

Modern color radar requires some caution in that seeing green, which tends to be thought of as light precipitation, can be in a CU that's just about to become a full-blown thunderstorm that will turn the radar color to red when the storm suddenly dumps its water. Remember, radar only shows rain, and that is not always the only place severe turbulence can be found.

☦ A Clue

When the weather services report actual thunderstorms, they give the direction from which the storm is moving and its velocity. This velocity should not be confused with the general speed of the weather system, such as a cold front. The speed is strictly of the individual storm. There is a definite correlation between this rate and the storm's severity. If it's over 20 kts, the storm will be strong, and if over 30 kts, it will be severe. In one of our worst airline thunderstorm accidents, the cells were traveling more than 60 kts. If a rain shower—not a thunderstorm—is moving over 50 kts, it will have strong turbulence and deserves caution.

✝ The Different Kinds

Now what do the different thunderstorms mean to us? Air-mass thunderstorms are generally scattered, and theoretically we see them and wander around, getting where we are going without flying through any storm.

We see this in the Far West, where thunderstorms occur in classic form. The visibility in the semiarid areas is excellent, and if anything less than 50 miles is reported, one looks at it with suspicion.

Here, it is easy to waltz gracefully around the big cumulonimbus, watching its dark, foreboding rain and flicks of lightning from a safe and interesting position. Away from these regions of unlimited visibility, however, air-mass thunderstorms become more of a problem as we fly in midwestern and eastern summer haze.

If we are working our way through air-mass thunderstorms in reduced visibility, the place to be is on top of the haze level, where we can see the thunderheads. It's worth repeating that when doing this it is important to make certain that broken to overcast clouds do not

Isolated thunderstorm. Note overhang on right; not the place to fly!
Hail often falls from there in the clean air.

Bright spots ahead, but very risky flying through those openings—go around!

creep in under us and that we keep track of the general weather situation.

Air-mass thunderstorms do occasionally line up in a sort of fake front if the condition favoring the thunderstorms is strong. This generally occurs in the late afternoon.

✝ How High

Thunderstorm tops are high—35,000 to 80,000 feet, depending on the part of the world. Except in rare situations, any CB worthy of the name doesn't stop at 25,000 feet. Almost any fully developed thunderstorm will keep right on climbing until it pokes up into the stratosphere.

The tops are generally lower in far northern and far southern latitudes, and so is the stratosphere. I've flown over the top of thunderstorms at 33,000 feet above the North Atlantic, but in southern Italy I've flown next to thunderstorms at 35,000 feet that looked as far above me as I was above the ground.

Thunderstorms' tops don't extend much into the stratosphere because the stratosphere is an inversion: the air is warmer than the rising storm's air, and so it's stable and shuts off a thunderstorm even though its momentum may drive it up into the stratosphere for a few thousand feet or so.

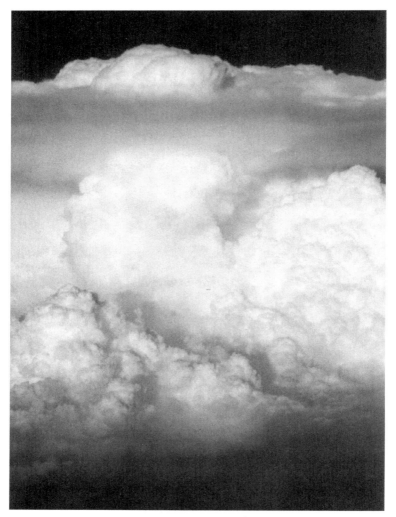

Thunderstorms—taken at 33,000 feet. Foolish to try and top them.

In the United States the tops are rarely below 35,000 feet. It's obvious that small aircraft don't try to top thunderstorms and big airplanes don't too often either. There are problems for both kinds.

If a pilot does try to top a thunderstorm, or a line of them, that pilot should be prepared for turbulence even in the clear; rough air often extends above the storm's top, and if an airplane is staggering along at a high angle of attack, struggling to get on top or stay there, it may be close to the thrust limit. If an upset or stall occurs there's not enough power to help recovery, and thousands of feet may be lost getting back flying. An upset with a jet, particularly, is a very tricky thing to recover from. The airplane generally pitches up first, and the natural reaction is to trim forward; then a dive follows and big control forces are needed to overcome the nose-down trim and dive. It's probably impossible! It's possible that the heavy pulling back, plus the speed, will load the stabilizer trim so it is locked. There's nothing to do except unload it by letting up on the hard pull and then trying to trim. This is one of those theoretical but highly impractical tricks. The upset is a complicated area having to do with speed of sound, shock waves, and air density. The recovery is often a matter of diving wildly toward the earth, sitting there pretty much helpless until the speed of sound changes with temperature increase and the air becomes more dense because of lower altitude and higher pressure, then the airplane gets enough control force for the pilot to recover. A number of these didos have occurred, and it's a testimony to the structural strength of jets that most held together, albeit some were stretched and strained.

It's impossible to overstress the danger of trying to top storms, and even going over them with apparently good speed and control is a more risky flying condition than one may realize.

✛ The Cloud Layers

It's not difficult to get in a bad situation trying to stay on top of lower clouds around storms. Let us say we are on top of the haze level. We see numerous cumulus (CUs) poking up through; they look like bunchy, thick cauliflowers. They are easy to wander around, but finally they seem to be connected by lower clouds and the top of these lower clouds approaches our cruising level. This can happen, depending on the stage of development, anywhere from 8,000 feet to more than 30,000 feet.

Almost without realizing it, we've been climbing to stay on top and our airplane is grunting to do the job. We are being suckered in, as the blunt expression goes.

Big cumulus tower to either side of our course. Ahead is a lower spot with bunched-up CU. We head for the lower place and hope we'll get over and beyond before it builds up to our level.

✛ They Grow Fast

Something we should realize, however, is that those CUs can build very fast and turn into thunderstorms. I watched one build in Texas. At 11:30 it was a pleasant-looking CU, and at 11:45 it was a big, solid, bunched-up-looking CU, and at 12:00 it towered to tremendous heights, its bottom was black, rain poured down, and lightning crashed to the ground!

When one stops to consider that thunderstorms in the CU stage can have updrafts of as much as 8,000 feet per minute, with 2,000 to 3,000 feet per minute fairly routine, it's obvious that a growing thunderstorm can outclimb an airplane of average performance.

So as we sneak through that "low" spot, we should realize that there's an excellent chance that the "low" spot will come up to us, around us, and envelop us. One cannot try to top growing CUs un-

Cumulus building. The low spot between will build, too, creating a situation in which one will have to climb higher and higher to stay on top. Thunderstorms will probably develop. For VFR pilots it's a situation that requires working calmly on getting down and out of it. For IFR it may mean an instrument clearance or climbing on top if the airplane can get high enough—15,000 feet plus—and then wiggling between thunderstorms.

less willing, able, and prepared to fly through a thunderstorm, be-cause that's what it may finally come to!

A VFR pilot flying on top of the haze level to spot CBs hasn't any business flying on top of clouds that are bunched-together masses, poking up to higher altitudes in places, with the ground only occa-sionally visible. When the CUs surround one, and the only path ahead is over some, or through low spots between them, then it's time to turn around or land.

The instrument pilot who feels it's okay to flirt with growing tops and try to slide over the "saddles" between them had best remember that no matter how experienced one is, a stalled airplane in a thun-derstorm is in a desperate situation, and the cleaner the airplane, the more desperate.

As we said, if this kind of flying is going to be done, then one must be willing and able to fly in a thunderstorm. What's willing and able? First, one must be a competent instrument pilot who can fly the airplane well in heavy turbulence. Then, one must have enough in-struments, good engine-heat capability, and radar.

☩ What's Inside All Those Clouds?

Let's go back to our line of towering CUs with the lower spot be-tween. As time goes on, the lower spots will be higher, and finally a solid line of massive CUs and CBs cuts across our course; they may be lined-up air-mass thunderstorms in the afternoon, or a genuine front. Either way it looks impressive and has the appearance of one big line of thunderstorms.

If we could really see inside that line, we'd find a row of heavy clouds with numerous thunderstorm cells inside. If we barged through and missed the cells, we'd get light-to-moderate turbulence and light-to-moderate rain and, on a chance basis, hail. If we barged through and encountered a cell, we'd have a wild ride, and if it's a cell just growing, young and vigorous, we'd have a terrible ride.

This has often been demonstrated by two airplanes going through a husky cold front on the same airway, same altitude, and perhaps only a half hour apart. One pilot says the ride was horrible, terrified all the way. The second pilot looks a little questioning and may think the first pilot an alarmist because the second airplane's ride had been pretty decent. The difference in rides wasn't due to a dif-ference in fortitude between the pilots; it was simply that one missed the cells and the other didn't.

✝ Radar

This brings us to radar. With it we have an excellent way of seeing the cells and avoiding them, although it takes learning, experience, and radar equipment in good condition. But we should never be on instruments when thunderstorms are nearby unless we have radar and know how to use it.

Radar is a tremendous aid in thunderstorm avoidance. What it does is let you see cells on a scope that you cannot see by eye because of other clouds and darkness. How does it do this? By the signal reflecting off rain, which says that radar isn't a cure-all because it really doesn't show all turbulence, only that associated with rain. There's lots of turbulence, often the most severe, outside the rain area! So it's showing you a "core" of problems surrounded by turbulence and trouble not shown, the extent of which you cannot discern. You have to "guess" how far away to stay. So schooling and experience is needed to get the most from radar; one just doesn't buy one, turn it on, and go fly thunderstorms. I strongly recommend that you attend one of the seminars available on how to use radar before you go out and try it.

The fact that radar bounces off rain also causes a problem in that it bounces off the rain of a cell ahead, for example, but doesn't get through that cell to reveal one just behind it; the signal is attenuated (weakened). What does that boil down to? Example: You are avoiding a cell the radar shows and go around it only to find that, when you turn the corner, there's another cell, or cells, facing you! Modern radars have tried to improve on this, but they still have a long way to go.

Look at the thunderstorms by radar as far in advance as possible, and then attempt to fly around the entire area, or the biggest area of cell congestion, rather than getting in close and using radar to squirm between cells. A good look and planning in advance can mean missing the entire mess.

There are more advantages to reading the picture in advance. For one, it allows time to tell ATC what kind of headings are desired to circumnavigate the area well in advance of arriving there. ATC will be in a much better position to cooperate and give the needed course changes.

Another point is that once in close to the cells, all will not be easy. New cells tend to generate close to old cells, so being in an area of cells, and close to them, means that more will develop and getting through it all without getting roughed up, even with radar, isn't going to be easy.

Being in close to the cells doesn't give much room to avoid them by the respectable distances that good procedures dictate; we'll talk about that in a moment. But all in all, the best way to use radar is to duck the entire area.

Remembering, then, that radar is a device to help you miss cells, it's apparent that when radar fails to show storms, it's a useless device. Radar is fallible; there are times when it doesn't do the job and we'll talk about them here. Before we do, however, let's make a point clear: If you are going to enter a thunderstorm area depending on radar to lead you through without getting in a cell, you had better be able and willing to fly through a cell if you have to.

Radar will take you places you'd never go without it. Suppose you get into this forbidding place and the radar fails; now what? You are either lucky and fly through without running into a cell, or you bang into one. If you hit one, let's hope you and your airplane can handle it!

Radar's important feature is that it reflects from rain, but at high altitudes the cloud doesn't have rain. It has snow and ice crystals instead. The radar beam doesn't reflect as well from this frozen stuff, and the cell becomes difficult to see on the scope.

The way to try to overcome this is to point the antenna down toward the rain area for a picture. The trouble is that the radar then bounces off the ground too, the picture becomes confused, and individual storms are difficult to see. It takes a lot of experience to read a scope under these conditions, and even with experience it doesn't always work.

Radar maintenance problems are not always just a simple matter of whether the radar works or not. There are degrees of radar maintenance, and this makes thunderstorm reading difficult and less clear.

It is necessary to get all the information possible from the manufacturer and learn to recognize when radar equipment is not putting out its best. There are various ways of doing this; one is to note how much "gain" the tube requires for a picture. If the knob has to be turned farther toward high "volume" than normal, things are getting weak and it's time to have the set checked.

Let's list radar's limitations:

1. Failure.
2. Equipment deterioration.
3. Attenuation.
4. Poor reflection from frozen particles.
5. Difficulty reading in mountainous terrain.
6. Pilot experience with radar.
7. Radar only shows rain areas.
8. Ice on radome causes false or poorly defined targets.

With these listed we can further list two rules:

1. Even with radar, don't go into cloud areas containing embedded thunderstorms if you can avoid them.

2. If in cloud or darkness and avoiding thunderstorms with radar, give them as wide a berth as possible. Don't just skim around the edge of cells.

One important point is the general guide that most airlines use in determining how to miss cells. It's based on temperature and altitude. They say to miss cells by the following amounts:

- When the temperature is above freezing—5 miles.
- Temperature below freezing—10 miles.
- At altitudes above 25,000 feet—20 miles.

These are minimums, and it's better to give the cells a wider berth if possible; 20 miles at all times is best.

Under the blow-off clouds on the downwind side of a thunderstorm is the place to stay away from. Even though it is clear and you can see around the storm to a happy storm-free area, it's a sucker area where hail is apt to be, and often clear-air turbulence as well. In a confused mass of cells and, of course, at night, we cannot see this area. The blow-off drifts a long way downstream, so a rule is: avoid a cell's downwind side, go around on the upwind side, and clear it by one nautical mile for each one knot of wind at the flight level. That may be a lot of miles, but it's worth them all.

Best is to go around on the upwind side and avoid the downwind blow-off area. But one should not just skim by the cell on the upwind side because there is wind shear in the area caused by a speeding up of the wind close to the cell, and that's turbulence. Give the cell wide berth, and remember the rule of one mile for each knot of wind at your level.

Again, remember that radar only shows rain areas, and even the new radars that boast turbulence detection by use of Doppler techniques are only telling you whether a rain area is turbulent or not—and how much. It isn't revealing anything about the turbulence outside the rain area, where some very tough stuff can lurk.

☨ Spherics

Before radar we avoided thunderstorms by experience, witchcraft, and tricks. We didn't avoid all of them because we couldn't "see" the cells, so we unknowingly flew through them, from time to time experiencing very rough rides, and we lost airplanes, too.

Coming up on a line of thunderstorms, we'd watch the lightning closely and try to avoid going into the mass of clouds where the lightning was "hot." It didn't always work because the lightning frequently seemed to cover the sky due to reflection within clouds and the general bombardment.

We also tried to tune our ADFs to a low frequency, away from any station, and then watch the needle, thinking it was pointing out thunderstorms and where not to go. Some claimed considerable success with this technique. My own success was of very low order and I never thought much of the idea.

We looked up to the tops, picked the lowest, and went through under that. For a time there was a theory that flying through the heaviest rain was the smoothest way, but that didn't work either. One thing we all agreed on was not to land or take off with a thunderstorm near or on the airport.

Paul Ryan of Ohio was flying his Cessna Skylane across one of the Carolinas when he tangled with a thunderstorm that almost did him in. The turbulence was wild and he felt lucky to get through it alive. Mr. Ryan, an electrical engineer, decided that something needed to be done about thunderstorms and avoidance for the single-engine "little guys" of aviation because most couldn't afford radar and even if they could, how could a radome and moving antenna be installed on a small single-engine aircraft?

He started, much as we did, with ADFs and eyes, to "see" the electrical discharges—lightning—that go with all thunderstorms. After much research and development he created what's now known as Stormscope. Far from our ADF attempts, Stormscope reads the electrical discharges, works them over in a computer, and displays them in front of the pilot. There is no movable antenna, or radome, rather just a small, fixed antenna and an instrument on the panel with a CRT presentation of dots that are electrical discharges, hence thunderstorms—or storms. Most models of the unit work through a 360-degree range so one can see the storms all around before takeoff, which can be very handy.

In flight the storms show up and can be avoided. The presentation is more widespread than radar and is less adaptable for squirming one's way through a line of thunderstorms—although radar often gets one in trouble trying to do this. As we said, radar should be used to avoid the area when at all possible. This is Stormscope's role. Since radar reveals the rain area and, one thinks, turbulence, it sometimes

*Stormscope display: The dots are lightning strokes, hence,
thunderstorms. There are many strokes in each storm, so the bunches
show the big action. As with any instrument, carefully review the
manual before use. Then work cautiously, gaining experience as you
go along. Strike-Finder has a display along the same lines.*

leads one around rain areas that do not contain turbulence. Stormscope shows the lightning, which says that the rain is a real thunderstorm and stay out. But if it isn't a thunderstorm, it will say that there's
little or no turbulence in that rain, awful though it may look. Numerous tales are told of pilots seeing dark, foreboding rain, but Stormscope not showing lightning. Brave souls have flown through these
areas and had good rides. This isn't recommended. Similar stories are
told of paralleling a front that looked mean, nasty, and impenetrable,
but Stormscope showed an area without electrical discharge and a
way though.

So where does it come out? Stormscope, no doubt, shows the location of thunderstorms, but not closely enough to weave through a
mass of them unless a clear, wide path is revealed. Radar shows more
detail within a mass of storms and, if you must work through them, it
helps, but with its treacherous limitations. Ideally one would like to
have both systems, and they are available merged together on one
scope. Some people have both, and the remark often heard is that

they use Stormscope more than radar. But it's there for all the single-engine and less expensive aircraft. And we can be thankful that Paul Ryan got himself scared in a thunderstorm—enough to do something about it.

Since Ryan's invention of Stormscope, a competitor called Strike Finder has come on scene. It claims to be able to do things Stormscope cannot; but looking them both over, the differences are not basic and both do the same thing, show where the lightning is.

The technique of spotting and displaying lightning has come to be called *spherics*—a word somehow derived from atmospherics. So we'll refer to these units as spherics systems and let it go at that.

✝ More about Air-mass Thunderstorms

We were talking about air-mass thunderstorms; it's important to realize that these come in several different kinds. The simplest is the convective kind. They are an eventual development from a thermal source, a hot spot on the earth. Obviously, when the sun goes down the heating does too, and the storm dissipates. If it doesn't, then the storm isn't just a convective storm; something else is creating it. The real thermal convective storm may take a while to die and doesn't disappear as soon as the sun goes down; a big bunch of clouds hangs on into the darkness with an occasional flick of lightning, but it's on its way out and one can see and feel that it isn't holding or increasing in strength.

Thermal convective storms are easy to run around when they are scattered or isolated. Sometimes, however, they aren't so scattered. Whether they are scattered or broken depends on how much moisture is in the air mass and how unstable it is. This is pretty difficult for a pilot to judge, except that forecasts give an idea simply by the number of storms they call for. Our old friend the temperature/dewpoint spread is another indicator; if the dewpoint is high, the air is moist and more likely to produce storms. You can "feel" this too, and since weather appraisal is inexact to begin with, one shouldn't laugh off old sayings like those about rheumatism that hurts when its going to rain. On a hot, muggy summer day, any weather-conscious person can sniff the likelihood of storms.

We are still basically animals, and despite years of sophisticated city living and separation from the primitive state, we have a feel for weather, feeling it in our bones, as the saying goes. Naturally we don't use instinct for our basic weather judgment. Science is the important thing, but that old instinct is worth using, particularly when it raises suspicions that all is not as good as a forecast may say it will be.

✛ A Cloud Base Hint

Incidentally, temperature/dewpoint can help in finding the base of cumulus clouds, or how high one will have to climb to get up out of the lower-level convection. Take the temperature/dewpoint difference in degrees C and multiply it by four hundred. The answer is the altitude in thousands of feet. For example: temperature 35, dewpoint 20. The difference is 15 degrees. This times 400 is 6,000. The base of the cumulus will be about 6,000 feet. (If dealing in degrees F, obtain temperature-dewpoint difference times 2, plus about 400 feet for the bases; temperature 93 degrees, dewpoint 67 degrees, difference 26 degrees, × 2 (add two zeros) = 52(00) plus 400 feet makes the bases about 5,600 feet.)

✛ Other Air-mass Thunderstorms

To sum up, air-mass thunderstorms are generally isolated and occur in the afternoon when heating has had a chance to work on them. They can also occur in overrunning air, not necessarily from a warm front, but from slopes acting like a warm front. This will carry thunderstorms into the night, but probably their base will be high.

Air-mass thunderstorms can bunch up and look like a front, but they can generally be flown around. In the parts of the world where visibility is good, such as in the far western United States, this is easy. In the more moist sections, where haze reduces visibility, it's more difficult; a pilot is best on top of the haze level where the CU buildups can be seen. It's important, of course, not to get trapped on top when doing this. It is also important not to try to top thunderstorms.

It's dangerous to sneak around underneath high, overhanging anvil clouds because that's where hail falls from. It falls more outside the storm than inside it.

It is dangerous to sneak under thunderstorms where visibility is reduced and there's a chance of flying abruptly into clouds, rain, and turbulence. Mountains may be buried in the haze too, and TV towers. Also, if a pilot runs into turbulence down very low, the situation is more hazardous than flight a bit higher, where there is room to wrestle the rough air.

While trying to get past a line of thunderstorms, down low we sometimes see the clear-cut black edge of the storm's base and beyond, happily, the pleasant sight of unlimited visibility and clear sky. We become eager to pass this line and go lower to stay "contact" and fly under. The nose is pushed down and the airspeed goes up, so that we're apt to start under the storm's edge going fast. But under this edge will be some really rough air with hard, sharp-edged gusts, and high speed will be dangerous. So patience is needed to keep slowed to turbulence-penetration speed as we sneak under and we should realize that there's tough flying yet to be done before our peaceful sky is won.

In early airline days, a DC3 was flown under such an area in a flat but high-speed dive by an overeager pilot. The jolts resulted in injury to 10 passengers and some wrinkles in the airplane!

✝ **Frontal Thunderstorms**

We visualize frontal thunderstorms as part of a dark, vicious cold front moving across the countryside with high winds, pouring rain, and much lightning, and that's a pretty accurate picture. But there are warm-front thunderstorms too and they are different.

What's the difference? Well, it's mostly in the height of the storm's base. A cold front's storm base will be close to the ground, with no room to fly under without getting the full force of the storm. But a warm front's storm's bases are generally aloft. How far aloft will depend on the frontal slope. I've seen thunderstorm bases as high as 16,000 feet, which is unusual. But I've seen lots at 4,000 and 6,000 feet. Between Chicago and Kansas City, for instance, I've often observed thunderstorms reported over most of the route. But by examining actual weather reports closely, one could see that the bases were high, which indicated a warm-front condition. A flight at 2,000 feet put the airplane in smooth air, although there were clouds, stratus from falling rain, and lots of rain and frightening lightning all the way.

✝ **The Surface Wind Tells**

How do we learn if it's that kind of condition? First, as always, is a look at the big picture. Where are the fronts and what kind are

A great picture of a line of thunderstorms along a cold front. The "open" places between the storms are very high and will tend to close up as other cells develop. This isn't a front to try VFR. It's an instrument flying job, with radar equipment required. (NOAA PHOTO)

they? Next, a careful study is made of the surface weather reports with particular attention to surface wind. If the storm bases are high, the surface winds will not be strong and gusty at stations reporting thunderstorms. The wind will pretty much stay normal for the general circulation. This is the best clue that the storm is a high-base one.

A warm front will also have storms over a large area with extensive cloud cover. A cold front, in contrast, will show up as a line of thunderstorms and can be seen on surface weather reports by the strong ground winds and a dramatic wind shift as it passes a station.

A warm front is of special interest to the turboprop and light pressurized twin because their normal altitudes may put them near the warm front surface aloft, where the thunderstorms are wildest. The pure jet will get on top of all this, or up where it's easy to dodge around, and the little guy will be lower and probably under the thunderstorm bases.

✝ How to Tell a Front's Toughness

The intensity of a cold front can be learned by studying the sharpness of the wind angle ahead of and behind the front. If the wind in advance of the front, for example, is from about 210 degrees and the winds behind the front are from 340 degrees, we can bet that the front will be a wild one. Also, the larger a low's warm sector, the more violent a front. The bigger the temperature difference between the warm and cold side of the front, the tougher it will be.

It is also important to check the speed of a cold front by looking back over the sequences to see how fast it has been moving. If it speeds up, you can bet it's getting stronger.

And, of course, there is our old friend orographic influence; if the front is moving up sloping terrain, that too will make it worse.

✝ Prefrontal Squall Lines

Cold fronts have a phenomenon associated with them that complicates the picture: the prefrontal line squall. This is a line of thunderstorms that breaks out 100 to 300 miles ahead of a cold front. They occur mostly in the afternoon, when heating adds to the zip of instability, along with the lifting effect of convergence in advance of the front. All this is difficult for the day-to-day pilot to analyze; simply always be suspicious of a prefrontal line squall when a cold front is on the move.

This squall line ahead of the front is where the toughest thunderstorms are found. It is where most tornadoes occur. It is where most hail is found. It is a good thing to avoid!

As a prefrontal line squall develops and intensifies, the cold front behind it tends to diminish and sometimes can be difficult to find on the weather sequences. As evening approaches, the prefrontal squall line will diminish and then the real cold front will regenerate and make itself known again. What this means is that if we negotiate a prefrontal squall line, we may think we have finished with its violence, but it may turn out that 300 miles along, as night approaches and we fly toward the dark, our cold front has perked up and we find ourselves facing another line of thunderstorms.

A good rule of thumb is that prefrontal line squalls are generally not active within 150 miles of a low center, nor more than 500 miles out from the center. They show up in an area 50 to 300 miles ahead of a front and roughly parallel to it. Sometimes the breaks in a prefrontal squall line are larger than in a solid cold-front line, but this is a flimsy thing to depend on.

✝ Some Rules

The same rules apply to cold fronts and prefrontal line squalls:

1. Flight on instruments should not be attempted unless equipped with radar in good working order, and with a knowledge of how to interpret what the radar shows. A look with a spherics system in advance will show how active the front is and if there are any clear places through it.

2. Low flight underneath should not be attempted because of extreme turbulence and the possibility of heavy hail.

3. The best procedures are to fly well around the line of storms or to land and wait for it to pass. Landing isn't too great an

inconvenience. First of all, these sharp lines of weather pass fairly quickly. Second, in a year of flying, most pilots don't encounter such conditions more than a few times. Of the total year's time flown, that rare couple of hours sitting relaxed on the ground is well worth it.

✝ If We Fly Through

Now, of course, comes the time when, for some hard-to-imagine reason, we barge ahead determined to go through that cold front. Where should we take it on?

Down low isn't good, as we've said before. Another place to avoid is the area where temperatures are −7 degrees C to +4 degrees C, (20 degrees to 35 degrees F.) Research has shown that this is the area where most lightning discharges occur. It's also an area where carburetor ice will be at its maximum.

A logical place to go through is just above the haze level, at the inversion. Here there is a minimum effect from heating, and that may be a place where the violence is a little less, and I emphasize *little*.

We can visualize flying toward a front. The sky is hazy and visibility poor in the hot mugginess. Cumulus clouds dot the sky. To avoid them, see better, and be in smooth air, we climb above the haze level. As we climb, the temperature decreases, but just as we come up out of the haze to where the visibility is unlimited, the temperature rises. That's the inversion.

Now we can see the CU sticking up through the haze. We are over Ohio flying at 10,000 feet. We should be listening to each weather broadcast on our route, noting which stations have southerly or northerly winds as an indication of the front's location. The wind velocity, the abruptness of the wind direction shift, and the temperature difference across the front are checked as well as the presence of thunderstorms at various stations en route.

Flying further west we note that the tops are climbing. We are at 12,000 feet and ahead is a line of CU from one end of the horizon to the other. That's our front.

This line of CU reaches very high. At first we have hopes of topping it, but experience says we're kidding ourselves. The line has higher places with anvil tops. These are the cells of the storms. With radar we can see them and begin to plan a route through. Without radar, we'll make an educated guess. Spherics would show the hottest electrical areas.

Close to the front there will be a towering cloud wall extending to neck-stretching heights. Below will be stratus and low, boiling roll clouds. At our level, just above the haze, we'll see some stratus and layers. The wall of the front is a solid mass, bulbous in places; its color is eerie, anything from misty white to black yellowish green, depending on the sun's position in relation to the front.

We study the complex picture before us and try to decide how to take it on. We don't want to descend and try to get under that roll cloud, because the air is wildly rough right there, probably the roughest part of the storm. It's tearing up in tremendous currents and it's chopped up and inconsistent. Close to the ground the wind over the surface adds to the thunderstorm's normal drafts. Progressing through this roll area, we'd run into heavy rain and zero visibility. The rain might affect our airspeed, and there are also localized pressure changes that will affect the altimeter's accuracy. (No one can tell you how much for certain, but some knowledgeable researchers think it may be as much as 1,000 feet!) So let's forget low down as a place to go through.

We look up, way up. We've probably studied this area for a long time as we approached the front. What we are looking for is the lowest point in the top of the long line of clouds, a saddle between towering CBs. When we find it, our first tendency is to climb right up there and slide over between the CBs. But as we've said, this is risky business because just as we are approaching the saddle, the clouds will build up in front of us and suddenly there isn't any saddle, or it's much higher than it looks. The risk, to repeat, is that we'll find ourselves struggling high up at a mushing, high angle of attack ready for a stall. Unless we can zip over this saddle between CBs with good speed and control, we shouldn't fool with it.

☩ At Night

Finding the right place at night is more difficult, but the flashes from lightning help. Approaching the storm area we study the cloud structure in the brief lighted-up periods as flashes occur. Often we can spot the individual storms by the concentrations of lightning. It is well to do this from as far away from the cloud mass as possible because close to the cloud the origin of the lightning becomes confused as it reflects and lights up all the clouds.

Lightning can be seen from a long way off at night, and storms that may seem on top of you can still be 50 or 100 miles away. Close to the storm area, when looking out, care must be taken that a nearby lightning stroke doesn't temporarily blind you. When you're that close in, it's time to have the cockpit lights up bright and the decision made as to where to penetrate the mess.

✢ Where to Bore in

We've decided that we cannot top the line anywhere. So we've picked a point under the lowest place in the top of the line confronting us. We're above the haze level and eye to eye with the solar plexus of the storm.

Before barging in we ought to study the color of the precipitation ahead. If it's white and fuzzy, with a gossamer look, it may be hail. If so, then we'd better wander up and down the line some more and find another spot that looks darker, like rain, and not whitish, like hail. Let's restate, before we go in, that we don't recommend it. We have our spot picked and are about to go in. Now the problem changes from weather judgment to flying technique.

✢ How to Fly It

First we prepare the airplane. We pull down on the safety belt and put on the shoulder harness. We make certain that loose items are fastened down and passengers well strapped in.

Put on heat for the pitot, carburetor, or jet inlets. Now establish the airplane's best rough-air penetration speed, and note what the power settings are for that speed and where the stabilizer-trim control is set. Note the airplane's heading.

But first let's talk a little about the penetration speed. Basically it's the slowest speed possible without danger of stalling. We want it slow so that the airplane will be in its best condition to take heavy gust loads. Too slow, however, has proven to be almost more dangerous than too fast. As we've said before, when airplanes come apart in severe turbulence, the problem begins with loss of control, and an easy way to lose control is to stall, or to get a wing down and into a spiral dive.

Clean, swept-wing airplanes are worse in this respect than slower, straight-wing airplanes. If they lose control, they gain speed quickly. A swept-wing airplane isn't as straightforward in its elevator-control system as a conventional airplane because the amount of elevator control available is dependent on the stabilizer setting. On certain jets one runs out of elevator control unless the stabilizer setting is correct. In effect the stabilizer is the elevator and the elevator is like a trim tab. A more conventional airplane doesn't run out of control; it just gets harder to pull or push.

It's best not to fly too slowly with any airplane. Most aircraft manuals recommend the proper speed for turbulence, and it will be something like 60 percent above stall. A 100-mile-per-hour stalling airplane may have a rough airspeed of 160 IAS. Jets may have higher speeds, but again, it's best to consult the airplane manual. Know your airplane's best turbulence penetration speed; get it from the manual.

The airspeed we've picked is important, but equally important is how we fly it. What we do not want to do is chase speed with the elevator control and the airspeed indicator. The key to rough-air flying is to maintain a level attitude and do it by flying the artificial horizon. Note where level position on the horizon bar is with the airplane stabilized. Then keep it there. We do this with gentle pressures and not big pushes and heaves. Airspeed will vary, often wildly, and if we attempt to chase it by pushing or pulling, we'll soon get into some pretty wild attitudes. The trim and power settings should not be moved to any large degree.

The proper pitch attitude, power setting, and stabilizer position are important to know in advance so one can set these up quickly and not fight an out-of-trim airplane while trying to settle down to the proper airspeed and power setting. These settings should be learned, as we've said, on a clear day by trying them for various altitudes and weight conditions. But the best insurance is to get down to turbulence-penetration speed, power setting, trim, and attitude before one enters a turbulent area. Of course, there are times when we hit turbulence unexpectedly; that's why the clear-air experimentation is useful to give us guidelines to shoot for when things are suddenly wild!

We have to face the fact that it is difficult, often impossible, and generally undesirable to maintain a given altitude when flying through the business part of a thunderstorm, ATC or no. Of course, we try to maintain the proper altitude, but only as long as it doesn't require appreciable power or attitude changes. Too diligent an attempt to maintain altitude will result in wild attitudes and/or power changes which lead down the path to loss of control.

☦ Are We Scared?

A thunderstorm line is a pretty awesome sight. It's scary, and any honest pilot, no matter how experienced, will tell you so. This fear is a factor to consider.

We've mentioned that scared or not, a weather-flying pilot has to control emotions and keep imagination subdued or used only to advantage. The pilot has to fly carefully and thoughtfully, even though scared to a knee-knocking state.

The toughest place to keep this fear under control is in a thunderstorm. It is dark, it is turbulent, rain comes down in a deluge, lightning flashes close, you can hear thunder sometimes, and occasionally ozone from a nearby lightning passage can be smelled. A lightning discharge may make a brilliant flash and loud bang as it goes off the airplane. An irritating, hashy noise may be tearing at our eardrums from static in the radio. It's a hell of a place for anyone to be! It takes a strong will to say, "I'll watch the horizon and keep it level. I'll hold that heading, keep the wings level, and it will all turn out okay." But that's it, and the only way to do it.

To help our situation, we should have the cockpit lights turned up full bright. If a lightning strike occurs, this will prevent temporary blindness. We should drop the seat as low as possible and keep our eyes inside the cockpit for the same reason. Bright lights and a low eye level help protect from a blinding flash, and they help psychologically, too. The bright cockpit lights keep us from seeing out, as does the lowered seat, and this helps shut out the terror. It puts a "protective" barrier between us and that awful world out there. It's a phony barrier, to be sure, but if it helps subdue panic, who cares?

So we come closer to the actual clouds. The first thing that makes us feel something's about to happen will be a strong updraft. It probably will be smooth but powerful. The airplane wants to go up, and if we push ahead to maintain altitude, the airspeed will go up. We can pull power to keep the speed low, but then the engines may cool so there isn't heat enough for carburetor or jet inlet anti-icing. If we don't pull power, then we get a nose-high attitude that's undesirable. We are probably best off by leaving the power set at our best rough-airspeed setting, keeping the horizon on the position of pitch for level flight, and the devil with altitude. If we are using an autopilot, the altitude hold should be off and the speed hold off too, if there is one. The human pilot should keep the nose on the horizon bar with the manual pitch trim. The autopilot will only be doing the heading and lateral work, which is a help and important, but it is stupid in its handling of altitude-attitude-airspeed relationships.

Autopilot or hand-flown, the wings should be held as level as possible, even if it takes big aileron movements and a lot of work. Nonlevel wings hurt speed control because an airplane wants to descend when it's banked, and level wings help to keep us going straight. The last thing we want to do in a thunderstorm is wander.

We start in with a heading that we feel is the shortest way through, and then we hold it tight to get through quickly.

The biggest updrafts occur outside the cloud. (Paul Bikel, a glider pilot, flew north along a front, in clear air, in the updraft for over 500 miles!) Rain hasn't started, but we don't like the feel of the updraft because we know that when it stops, there's going to be a jolt. And just as we start into the cloud, our updraft stops with a jolt and the air becomes very turbulent. Side gusts, downdrafts, updrafts...it's wild. It's dark and as the airplane is jolted and jarred, heaved up and squashed down, the rain begins with slaps of big drops at first and then, suddenly, inundation. The rain flows over the windshield like a river, and it is noisy. Most people would say it is hail. There may be soft hail and it may make a frightening noise, but it isn't big, solid hail. Solid hail is a noise to end noises. It's unforgettable, tremendous, and no one ever mistakes it for heavy rain once they've heard it.

And what if it is hail? There isn't much to do. Keep a firmer grip on emotions, but above all hold the heading! Hail is confined to a small area. Turning back will probably result in a longer time in the hail. Also, a turn will expose the side of the airplane and break out windows, and the amount of up-elevator needed for the turn may expose enough fabric, if it's a fabric-covered airplane, to tear the fabric.

I saw a DC3 once that had flown into heavy hail in Kansas, and the pilot had turned around. I looked at the airplane as it sat in the hangar, a battle-scarred wreck. The leading edges were beaten in, the landing lights broken out, and the windshield, too. But most interesting was that every window on one side was broken! Fortunately, it was a cargo flight without passengers. The fabric-covered elevators had numerous rips and slashes. So don't turn around in hail.

There are some interesting points about hail. Most of it in the United States occurs between the Mississippi River and the Continental Divide. April, May, and June are the worst months. The worst time of day is between 2 P.M. and 10 P.M. Cold-front and prefrontal thunderstorms are more apt to have hail than air-mass thunderstorms.

Hail forms in the storm's building stage and falls when it's in the mature stage. It is largest near the freezing level. Flight well above or well below the 0-degree point decreases the risk.

Looking again at the tropopause and thunderstorm heights will give a clue to the possibility of hail. If the radar reported that storm tops push up above the trop by 5,000 feet, one can expect to find at least half-inch hail.

In the black heavy rain, the turbulence keeps up, but there is a feeling that it is diminishing. It's still wild and rough, but we feel more in control of the situation. There was probably a point in the early part when it was so rough that we doubted our ability to keep control if it got any worse. Now, in heavy rain, we feel that the out-of-control threshold has been lowered in our favor. The rain does dampen the turbulence to some degree.

✝ Something to Be Said for Rain

There was a time, before radar, when pilots said that if you must go through a thunderstorm pick the blackest part with the most rain. When radar arrived this became taboo and we learned the error. The reason, of course, is that radar shows where a cell is by reflecting from the rain. So radar says, "There's the heavy rain, that's the storm's cell. Avoid it!" This is essentially correct. But the most severe turbulence isn't exactly in the middle of the rain. It's close by, however, and if you miss the rain area by a good margin, you'll probably miss the worst of the turbulence.

Don't be sold on the idea that the center of the rain is the wildest part of the storm, although it may have the most violent downdrafts, or bursts as they're called today. The roughest place is probably in the area just ahead of the storm, near the roll clouds before you go on instruments.

✝ Fly!

Now back to the storm. Our job is a simple one, really: just fly the airplane. Fly attitude; keep it level and under control. We need a certain attention to the engines to make certain the carburetor heat is sufficient and that they are not icing. There's little navigation to do. We can glance at an omni and try to keep on course, but we shouldn't do any excessive squirming around with large bank angles. It isn't going to be far enough through the roughest part to get much off course.

✝ Electrical Discharge

Lightning will flash in this darkness, and some of it seems close. The sky is lit up for moments, and its brightness is scary. Between flashes, when it's dark, we can see small flicks of fire dance across the windshield, and a quick peek at the propellers, if we have them, shows a neonlike band circling their tips. This is Saint Elmo's fire, more formally called *corona discharge*. It means that the airplane has been flying through electrical fields, absorbed energy, and is charged. If it collects enough charge and passes through an area in the storm that has a big charge of opposite polarity, the energy may jump between airplane and cloud in a type of lightning.

It goes off with a loud banging PLOP and a brilliant flash of light. If you were looking out you'd be temporarily blinded. But looking in or out, it will scare you.

It's over in a second and we are surprised to note that the airplane still flies, and the wings haven't fallen off! If we could see the damage we'd probably find a half-dollar-size hole near the wingtip or at the tail cone or some other small-radius area. We might find that a radio has been damaged and we should look with suspicion at the magnetic compass. Beyond this we are okay. Thousands of these discharges have occurred with few serious results, except as noted in the beginning of this chapter.

These are commonly called lightning strikes. They are the only kind an airplane gets. Airplanes aren't "hit" by lightning as we visualize a person on the ground being hit. There is always an electrical discharge between airplane and cloud. A person on the ground, actually, isn't really "hit" either, but is a point where electrical potential jumps from earth to cloud through the person. Though we always think of being "struck by lightning," the person "struck" is a part of the process of discharge.

The chances of a discharge can be lessened, as we said, by staying out of the region 4 degrees C either side of freezing. The majority of discharges happen where the precipitation is rain mixed with wet snow. There are little tricks, too, that may help some; they certainly will not do any harm. One is to turn on propeller alcohol if available; another is to flick the mike button now and then. What these two things may do is to help carry away part of the energy that's been built up on the airplane. It's a slim possibility, but may just be enough to keep the charge below the critical point. The key thing, however, is to stay away from the freezing-temperature area. The faster the airplane, incidentally, the more likely a discharge. Also, while keeping away from the freezing level is best, it's not a guarantee. The discharge I had out of London that resulted in the big piece being torn out of the 707 radome occurred between two cumuli whose tops were only 11,000 feet and the temperature was 14 degrees C.

The electrical charges come in two ways, and they are worth talking about. One we have explained: the accumulation of a charge by flying through electrical fields. These are huge, with tremendous voltages, and are a part of the instability of the thunderstorm process. I always visualize them as big waterfalls of electrical energy; you fly through them and the airplane is covered by the energy and absorbs a great quantity of it.

The other kind of charge arises from flying through precipitation. This, crudely, is a friction process, like the crackling one hears when combing one's hair with a plastic comb on a dry day. This type of charging doesn't affect things nearly as much as the field type.

It's easy to see that when flying through a thunderstorm we get the field-type charge, and if we are flying in the wet-snow area we also get the friction type in its worst form; so we are getting the maximum. Getting away from wet snow deducts part of the charge— unfortunately the smaller part.

⊥ Static and Radio

All this charging causes static, which knocks out radio reception on certain frequencies. Fortunately, it rarely affects reception in the VHF ranges. It offends most seriously in HF frequencies, and in the ADF 200 to 400 kilohertz range.

An electrical charge looks for the easiest way off the airplane. This will be any small area, like a copper antenna wire. Of course, when the charge bleeds off an antenna, it makes a terrible noise in the radio. To prevent this a large wire with a polyethylene coating is used for an antenna, the entire thing being about the diameter of a pencil. Also, the insulators and attachment hardware are large and smooth so that no wire end sticks out in the air. The big wire and smooth insulators discourage discharge.

The other part of antistatic hardware is the little wicks we see sticking out from wing and stabilizer tips, and sometimes from the rudder and tail cone also. These are designed as a place where the charge can bleed off quietly and easily.

All this helps, but only for the "friction" type charging; it doesn't impress the electrical-field type at all because it is so big the wicks cannot bleed it off, nor can large-diameter wire antennas contain it. If they could, we'd never have a big-bang discharge.

This was a major problem in the radio-range and HF days. It is much, much reduced now because of VHF, but electrical fields are so strong, even VHF will start to howl and squeal in thunderstorm conditions at times. It generally doesn't last long, but when one can hear it, a discharge may occur! For long-range airplanes needing HF for communication and aircraft equipped with Loran, precipitation-static discharge can be a problem.

The ADF will be knocked out by precipitation static, and it's important to keep in mind that ADF bearings will be very unreliable. Following an ADF indication could easily take one off course and perhaps set up a dangerous situation, particularly in mountainous areas. Look at all ADF bearings with much suspicion when in static.

☨ The Noise Is Annoying

While we are tossing around in the dark, wet, and rough inside of a storm, we can use radio noise in a couple of ways. If we have HF frequencies, or are using low frequencies, the noise will be so continuous and loud that nothing else will be heard. All the radio is doing then is aggravating our condition of apprehension. The best thing to do is to turn it off. It's surprising how calm it is, even in the turbulence, to suddenly have that infernal noise go away.

If we are using VHF, we leave it on. The occasional squeaking noise we hear may tell us a discharge is close. While at this point we cannot do much about it, it will at least have us prepared and not so startled when the bang and flash occur.

☨ Almost through the Storm

As we continue, the rain slackens, the turbulence quiets down, and patches of lightness bring encouragement. We may fly out of the storm quite quickly and dramatically find ourselves in brilliant sunshine. While we breathe a sigh of relief, it's a good idea to remain fastened down and alert until well away from the area.

We have traversed this storm from front to back. The worst part was first because the roughest part is the front, or leading part, of a storm.

Had we been coming the other way, the sequence would be reversed. The rain would begin, and turbulence start; it would become darker and darker and the rain and turbulence heavier. Toward the front side, light spots would appear, but it is then that we'd get the real rough stuff, that wild updraft and turbulence even after we have broken out of the clouds. Flying back to front, we want to be well through and ahead of the front before we relax.

Now suppose we are flying a jet and hit all this up high, at 25,000 or 30,000; what's up there? Lots of snow, for one thing, and turbulence, probably not as rough as down lower, but still very rough. There is more chance for hail, and electrical discharge—lightning—is quite possible. The massive amounts of snow, some of it in the form of very large flakes, may cause icing, especially inside warmer intakes where the snow is melted only to refreeze later in a colder area.

While doing thunderstorm research, I barged into one at 35,000 feet in a B-17. It was impressively rough, but most impressive was the size of the precipitation: The snow looked like big snowballs being thrown at us. This stuff got into the intakes, clogged them up, and, by restricting the airflow, caused supercharger ducts to squeeze in. The result was four very weak and helpless engines and a big, not-so-good glider! Fortunately, the high altitude made it possible to glide away from the storm, get lower, and reorganize things so that we could limp to an airport.

Up high or down low a thunderstorm is potent. Even over the top of one, turbulence will extend upward into the clear air a few thousand feet.

✝ Warm-front Thunderstorms

When flying warm fronts, we will be on instruments most of the time and it will be difficult to see individual storms. Occasionally the airplane will break out between layers, and then it may be possible to see the CBs towering up from the bottom layer into the higher layer.

Because we cannot tell without radar where the storms are in a warm front, we should fly prepared to get in one any time. And while they may seem a little less violent, they are still thunderstorms and can be as rough as any.

✝ Low Down

The best rule, in warm-front conditions, is to fly low. Keep down, near the minimum instrument altitude. Since we are dealing with turbulence, it's best not to fly *too* low, and I personally prefer 2,000 feet above minimum instrument altitude. This is dependent on terrain. If it's all flat, I'd feel secure 2,000 feet above the ground, but in mountainous areas I'd want more.

This low flying in warm-front thunderstorms is due to the fact that warm-front thunderstorms generally have high bases. The air has to crawl up the warm front to have enough lifting to set off the thunderstorm.

If we are flying a jet, we don't fly at 2,000 or 4,000, although I've known of this being done with considerable success—if fuel isn't a problem. With its high-altitude capability, a jet will sometimes top an entire warm front—over the ocean most certainly. But if a warm front with thunderstorms cannot be topped, then it's best to come down to an altitude where the airplane is flying well and to use radar to go through. The best altitude will depend on the jet's performance; generally speaking, about 33,000 feet is good. At high altitude, the storms seem more widely spaced and easier to weave through.

✛ Thunderstorms at Landing

Sometimes we are faced with the problem of our destination being covered by a thunderstorm, or having them so close that they affect our landing. What do we do?

First, we must approach the area and dodge any cells near the airport. We can do this by VFR procedures if we can see, or by radar. The complication, of course, is traffic control. If we wander around dodging storms, we have to keep ATC advised and get their prior okay. It's only logical that two airplanes from opposite directions might be headed for the same good area. A deviation also might put us on another airway.

ATC will sometimes help lead one through a thunderstorm area. This can be a mixed blessing. The bad part is that air-traffic radar does not clearly show thunderstorms. The equipment is designed to show airplanes and not weather. The storms show in a vague way on the scopes and some controllers feel they can use this limited picture to lead you through. But they cannot do so adequately unless there is a special weather radar nearby.

The best plan is to ask the controller if the radar has weather capability. If it doesn't, you are probably better off doing it yourself. The FAA is trying to cure this problem, and I wish them speedy good luck, but don't count on ATC as your method of thunderstorm avoidance. That's not their job, they are not always good at it (the busier they are and the worse the conditions, the less likely they can help), and their radar isn't designed for it. Use onboard radar and/or spherics plus much discretion. And don't let ATC talk you into flying too close to cells or tell you it's only a VIP 2, or it's not bad because some other airplane recently flew through it. They don't read storms that well, and as I've said, VIP levels can change quickly, as can the storm since the last person flew through the area. You be the judge—and boss!

✝ Don't Race Thunderstorms

A real hazard is a pilot trying to beat a thunderstorm to the airport. There's nothing wrong with trying to get there first, but there are a couple of possible pitfalls. One is diving and allowing the airspeed to get too high, and then suddenly flying into the rough area just ahead of the storm at this high speed. Another is racing to get to the field and then landing as the storm arrives, or even when it's five or ten miles from the airport. The wind shifts abruptly with gusty force, and a cross tailwind suddenly makes landing difficult, perhaps impossible. This, again, is where severe shear occurs.

With the wind shift comes heavy rain that obstructs all vision, and one may suddenly go completely on instruments—zero-zero—anywhere along in the approach. At 100 feet, for example, the pilot will suddenly be unable to see the ground. Even if the rain arrives just at touchdown, the visibility will be so poor that it will take luck to make a straight runout and stay on the runway. It's a terrific shock to suddenly learn how bad visibility is in heavy rain. Braking will be poor, possibly nil, on the wet runway. The airplane can aquaplane and not stop. With a strong crosswind it can also slide off the runway sideways. And always remember that thunderstorms on or near airports is the classic recipe for shear accidents.

✝ How to Turn Around

Now, an important point. Let's say we are racing a line of thunderstorms to an airport. The line arrives just as we do. We decided to pull up and abort because landing is impossible. Now, which way shall we turn, right or left? In the northern hemisphere it's best to turn right. Why?

The wind has probably changed to NW. Making a left turn, we will have a momentary drop in airspeed, bringing us closer to stall; the sudden new side component of wind will also tend to overbank the airplane; an intense downdraft will be present as the cold air arrives.

It all spells an airspeed loss that's difficult to regain. There have been accidents that were caused by loss of control when turning away from a thunderstorm.

Of course, if you are still in the southerly wind part of the approaching storm, you'd best turn left. The point we are trying to make is that if you get in close to a storm, particularly where the rain has started and the wind shifted, and then decide to turn around and get out, it is best to make a right turn.

There isn't always time or circumstances to do the correct thing, but right or left, remember the air will be gusty and shifty and it's very important to maintain enough airspeed above stall to take care of downdrafts and wind shears. If power is needed to accelerate the airplane, this isn't a time to be bashful about the amount. Pour it on, lots of it. It takes an airplane longer than you think to gain airspeed by power alone when pulling out of a low-airspeed condition.

✝ After We've Turned Around

When we have gotten turned around and away from the storm, what then? If it's an air-mass storm we may want to hold out in a clear area until it leaves our airport, and then go in and land. If it's a front, we'll have to pass through it before we can get to our airport. This case calls for a retreat to some still-clear airport where we can land and wait until the front has passed. Then we fire up and complete the flight calmly and pleasantly.

In summary, the first thing in any airplane at any time is to stay out of thunderstorms! If you must fly through fronts or heavy cloud cover that contains hidden thunderstorms, don't do it without radar and/or spherics. If the radar fails while you're in the middle of the mess or for some other reason you get caught, fly the best speed, fly attitude, and hold a heading. Remember power settings and stabilizer settings so that if excessive dives, climbs, or upsets occur, you'll know where things ought to be reset! Don't fly very low or very high! And most important, keep emotion under control and fly smoothly, and coolly.

15

Ice and Flying It

What's it like when an airplane picks up ice? It begins by forming on little corners of the windshield, and on the wing we see a polished look on the leading edge if it's clear ice, or a fine line of white if it's rime. If the windshield isn't heated, a smear of ice will cover it. Ice will begin to accumulate on pieces of the cowling, masts that stick out, and other un-deiced places.

Before getting into ice—when icing conditions are possible—an aware pilot knows what the IAS is and the right power settings, piston, or jet. Then, once in ice, these settings are observed carefully for any changes. Change in IAS is a sure way to know the ice is affecting the aircraft, and power changes say heat had better be applied quickly.

But the all important point is that at this early stage, right now, a pilot should work to get out of the condition! Don't cruise along hoping for better, don't grind around in a holding pattern, let ATC know immediately you want out! Get a different altitude. Go down if there are above-freezing temperatures below. Get on top if you know where the tops are.

Be careful about climbing, however, if you don't know what's up there. If it's a warm-front condition, you can climb into worse conditions, and climbing along the slope of a warm front is really asking for a load of ice. As a rough rule, we climb in a cold front and descend in a warm front, but this is variable. Each weather condition should be analyzed and studied by the pilot before deciding what to do.

Unfortunately we cannot make hard-and-fast rules about each weather condition. We know things in a general sense, but we have to look at the flight as an entity with all the parts that may affect the outcome: preflight weather study, fuel remaining, terrain below, availability of places to land—the entire jigsaw puzzle of flight, with each part influencing the decision of what to do.

Fortunately, fronts take up a relatively small portion of our weather time. The lesser weather, such as air-mass conditions, makes up most of it. We can come closer to rules for flying in these conditions, and they are covered in various places in this book.

But let's imagine we got into ice and didn't do anything about it, but instead just sat there and let it grow. We would begin to see an airspeed drop. This is always a little alarming. A couple of knots doesn't impress us much. Five knots begins to, and when ten knots have gone we get edgy.

The thing to do is pour on more power. We want to keep the airplane at as low an angle of attack as possible. We don't want to get in a tail-draggy situation with higher drag and ice forming back under the wing.

About the time our airspeed loss is bothering us, we develop a bad vibration. It is a hunk of ice breaking loose from the propeller. A little more propeller alcohol is needed if we have it, or more heat.

Now another vibration begins. The unheated windshield gets a thicker coating and we can't see ahead at all. On the corner of the windshield a big hunk of ice has built straight ahead into the airstream, and it's four to six inches thick! The vibration gets worse and now a howling starts. The howling is probably an antenna mast vibrating from ice disturbing the airflow around it.

The wings have a good coating, and if we have rubber pulsating boots we turn them on. The ice breaks off in chunks but leaves pieces that sit in the airstream and go up and down with the boots. We turn them off.

The engine cowls now have a large coating of ice on their leading edges. The scoops have ice around their openings. The propellers periodically heave off hunks of ice and the vibration increases. The antenna mast howls and shakes.

The wing has more ice and we try the boots again. Hunks break off, but some hunks don't. There's a messy collection of pieces of ice behind the boots. The airspeed has dropped more and we apply more power, resetting the carburetor heat as we do. The throttle is open an alarming amount. The ice on the engine cowls is almost out to the props. The vibrating and howling from masts is eerie, and then it suddenly stops. So does one of our radios. The mast has carried away.

The wing boots are only doing a partial job, and bigger hunks of ice go up and down with the boots. The airspeed continues to decrease and the throttle is wide open. The only way we can maintain speed is by losing altitude. The only way out of this mess is down, and we hope there are above-freezing temperatures down there before we

A radio mast, about 12 inches high, loaded with drag-causing ice. Things like this can't be deiced. The more of these "ice catchers" on an airplane, the less time one can stay in ice— deicer-equipped or not. (NOAA PHOTO)

get to the ground. The situation is pretty desperate but unnecessary if we had done something about getting out of the ice when that little bit first formed on the windshield corner.

Ice affects the flying qualities and characteristics of an airplane. The most serious thing it does is destroy smooth flow and make a different airplane of the one we know. The weight of ice is of secondary importance.

Ice affects the wing section. It affects the propeller in the same fashion. It collects on things sticking out that create parasite drag like antenna masts, wires, and cowlings, and makes them drag more.

Ice will also cause an object like an antenna mast to vibrate and howl severely and finally to break off, or half off, which may be worse. The first experience I ever had with ice was in a Pitcairn Mailwing biplane. Cockily, with a newly learned ability to fly instruments, I barged into a large midwinter CU. In a few moments I had an appreciable load of ice. It collected on the wires which held the wings on. The wires started to vibrate wildly up and down at least a foot each way. I expected them to break at any instant. I didn't have a parachute. I quickly flew out of the CU, the ice shook off, and the wires didn't break. I had become much older and wiser.

My next serious ice experience involved flying a Constellation across the North Atlantic. At 18,000 feet I encountered light ice, which covered a long antenna that went from a mast up front to a fin at the back. The antenna vibrated and finally broke near the fin. The long wire, with a big antistatic insulator on the end, whipped around beating the fuselage. I tried different airspeeds in the hope that the wire would "fly" out straight and stop beating the fuselage. Nothing helped and finally the insulator gave a sound whack to a window and broke the outer pane; fortunately there were two. But this meant I had to reduce the cabin pressure and descend. I descended into a worse icing condition. It was a moment of relief when we finally landed at Gander, Newfoundland.

The point is that often the gadgets on the airplane are the worst offenders, and if a pilot expects to fly a lot of weather the airplane being flown should be as clean as possible, with a minimum amount of stuff sticking out.

Flying a 747 I've had little ice and none of the howling, vibrating kind from gadgets that collect ice. The jets are "clean" airplanes and the "cleanness" reflects in their resistance to icing. There's also a temperature rise due to ram at their high speed, but that's another story and not very important here.

It is important to realize that deicing equipment does not allow one to fly indefinitely in ice. It cannot do this job! It will help and give one time to work one's way out of icing situations, but it will not allow one to sit there all day long. There are a lot of reasons why it isn't good enough, and one of the main ones is that deicing equipment doesn't cover the entire airplane. Another is that it doesn't completely clean off all ice.

So we have an important first rule: When ice is encountered, immediately start working to get out of it. Generally this means a different altitude, after a request to ATC. Unless the condition is freezing rain, it rarely requires fast action and certainly never panic action; but it does call for immediate, positive action.

Now let's talk about ice in detail. First, there are two kinds: clear and rime. They are what they sound like. Clear ice is a smooth, hugging type that is tough. Rime is crystal-like and pretty. It often forms in weird shapes and sometimes sticks out as long cones into the airstream ahead when it forms in windshield corners or other places where airflow patterns change. Rime ice breaks off fairly easily with deicer boots that pump up and down.

Clear ice doesn't, and one description of the difference between rime and clear is that clear ice cannot be removed by deicers and rime can. This is an oversimplification, however.

Ice comes in three classes: light, moderate, and severe. These are difficult classifications, which depend on the pilot's judgment; one person may think a certain degree of icing is light, while another calls it severe.

Generally speaking, the categories refer to the rate at which ice forms. In flying through freezing rain, ice accumulates very quickly and is called severe. Since it's rain, it has to be falling from an area warm enough to have water. The cue in freezing rain is to go up. If our equipment and our ability is limited, then it's best to quickly turn back or land. At any rate, we must do something fast because it forms fast!

I remember a DC2, flown by a friend, approaching Chicago's Midway. At McCool he ran into freezing rain. The field was so close he decided to make a run for it. He landed with almost full power dragging him in. There was so much ice on the airplane that it was necessary to chip it off the fuselage to get the door open! A few more minutes and he would have spun in. This was during the time, before we knew better, when we thought a DC2 could fly through *any* weather. We learned differently.

Light ice would be found in a thin stratus cloud in very cold air that doesn't have much moisture. Light ice can, of course, become moderate if we sit in it long enough. We can continue our flight in light ice, but not unless there is a quick way to get out of it should it become moderate, that is begin to form faster. There should be ceiling below so we can duck under the clouds, or a known and easy to reach top, or an easy to reach better weather area behind us.

The speed with which ice forms is the thing that counts because it affects what we're going to do about getting out of it. If it's light we can move slower, talk it over with ATC, and have time to work. If it's severe, however, we might be in an awful hurry, and just tell ATC what we are doing and that it's an emergency.

Before we go any further let's talk about deicers. What do we deice? Pitot heads of course, carburetors, or in the case of a jet engine, the engine inlet cowl, guide vanes, etc. This is obviously important because if the engine doesn't run, we don't fly.

Carburetor heat can be tricky. First, one should have a carburetor-air-temperature gauge, and then know exactly where the temperature bulb is located in the carburetor system. The reason is this: If the bulb is located in the coldest part of the carburetor, one only needs to pull on heat until the temperature gauge reads something over freezing, like 35 degrees F. If the bulb is located in the scoop, it may be necessary to carry as much as 85 degrees indicated to be certain that the coldest part, farther downstream, is above freezing. There is about a 30 to 40 degree temperature drop from the air scoop to the coldest part of the carburetor.

This can lead to trouble if there is not sufficient heat. A pilot might be flying in a condition of cold dry snow, for example. Generally this condition doesn't cause carburetor ice, because the cold ice crystals zip right through the engine and don't get up to a slushy, freezing stage. But if, in this very cold air, a person puts on heat as a precaution until a bulb in the air scoop reads just above freezing, the temperature would raise just enough so the ice crystals would begin to melt and then refreeze in the colder part of the carburetor system causing ice and trouble. I had a double engine serious power loss in a DC-2 learning about this, but fortunately I had enough altitude to get them going again by leaning and backfiring the engines to break the ice loose.

It's therefore important to learn where the carburetor-temperature bulb is located and how much temperature drop there is through your system, and then to use carburetor heat accordingly, always having it above freezing in the coldest part of the carburetor if conditions require heat.

Under most conditions it's obvious if heat is needed. It's used, for instance, during wet snow, in rain, in cloud with a near-freezing temperature, even in clear air with high moisture content if the carburetor is a sensitive one.

A sunny summer day may produce carburetor ice because warm air can hold a lot of moisture; so if it's warm and humid, carburetor ice is a real possibility, especially at low power, such as when idling during the glide on an approach to landing.

But sometimes it is difficult to know whether heat is needed or not. A close watch on engine performance will tell.

If the airplane has a fixed pitch propeller, watch the RPM carefully to see if it's decreasing. (To do this a pilot has to know the RPM previously set and also be certain the airplane isn't climbing, because even a small climb angle will decrease the RPM.)

If we get an RPM drop, it's a good indication of ice. Pull on carburetor heat. The RPM will drop further because the mixture has been upset with the addition of heat. Leave the heat on for a minute and then take it all off. The RPM will bounce back to its original value if the ice has been melted out. This tells, of course, that ice was the problem.

Naturally, we don't want to operate with this off-on procedure; so now, knowing there is ice, we pull on enough heat to get our carburetor temperature above freezing in the coldest part of the carburetor.

The RPM will drop when we do this because the mixture has become rich. The hot air we are now using is less dense, so there's effectively less air in the mixture. We lean the engine with the mixture control, and when we do so, most of the RPM loss should come back.

If we haven't a carburetor-air-temperature indicator, then we put on heat by guesswork, lean out, and keep a close eye on the RPM. If it holds, we've got enough heat. If another drop occurs, we should pull on full heat to clean out the ice, and then apply heat again, except this time a little more.

We don't want too much heat; we ought to try and use only just enough. Generally, if too much heat is used, leaning the mixture will not bring back the original RPM. If excessive heat is used, the engine may run rough and lose power.

If our propeller is the constant-speed type, we'll never notice any RPM change because as the engine loses power, the propeller pitch changes—gets flatter—and keeps the same RPM.

The way we tell, then, is by watching the manifold-pressure gauge for a drop. That will tell us there is ice. Then we go through the same procedure as mentioned before except that we use the manifold-pressure gauge instead of the tachometer to judge power loss or gain.

If we change altitude, we'll have to reset the heat-mixture relationship once leveled off at the new altitude. Most engines use rich mixture for climb, too. With high-powered outputs, as at takeoff, carburetor heat isn't needed, and if one used it on takeoff, there would be a big power loss, and even a chance of engine damage. So we clean out the ice before takeoff, take off without heat, and once in the air watch closely for any signs of engine icing.

During climb most engines do not use heat, but under severe conditions it might be needed. Any deviation from normal climb performance, on instruments, is a hint of carburetor ice.

We hear a lot about nonicing carburetors. These are injection types and don't have a carburetor venturi with its temperature reducing action. These nonicing engines don't ice very easily—they seldom do at all, in fact—but icing can nevertheless occur in them because it is still necessary to bring air in from the outside for the engine to breathe. If this air has the proper moisture content and temperature, freezing may occur somewhere in the induction system.

Clever duct design has reduced this hazard to a minimum, but icing can and does occur. A pilot still has to keep the possibility in mind. Because advertisements say "nonicing," we should not go along innocently believing it 100 percent true!

In the awful condition where there isn't enough heat to clear out the ice, a desperate trick might help. The trick is to try to make the engine backfire in the hope that the backfire will clear out the ice. Generally this can be done by leaning the mixture until the engine runs rough and backfires. If it's successful, the engine will come back in with a great roar as you enrich the mixture again. I've done it twice under desperate conditions, once, as mentioned, in a DC2 and the other time while flying a little Culver Cadet in wet snow. Both times it worked.

Jet engines are a much simpler matter. There isn't any carburetor and so there isn't any carburetor ice. But the air inlets, cowling, and guide vanes can collect ice just as a wing does, and sometimes even when the wing doesn't, because there is a temperature drop in that big, venturilike cowling that can cause ice occasionally, even in clear air if the air is very moist and the temperature just right.

Ice affects jet engines seriously as the inlet airflow is disturbed and cut down. Fortunately, the engines are equipped with hot-air passages within the cowls, guide vanes, and so on. The pilot simply flips a switch and hot air from the engine is routed through the passages to keep them warm. The only effect on operation may be a slight power reduction from air loss in the engine.

Jet inlets may form ice on the ground during a long ground hold waiting for takeoff during high-moisture, lowish-temperature conditions. The jet inlet causes an air temperature drop and ice may form on its walls in the above conditions. The airflow into a turbine engine is very critical; a relatively small amount of ice can disrupt it to the point that when we're finally cleared for takeoff and advance the throttles, we may get irregular engine action or a compressor stall.

So while holding on the ground, it's wise to use cowl heat periodically to prevent ice accumulation if the temperature-moisture setup is conducive to ice formation. There doesn't have to be visible moisture such as rain or snow, but simply high humidity; cold fog would be suspect, too.

The main point is to be certain that heat is used whenever there is a possibility of icing. All engines, piston or jet, get their deicing power from heat generated by the engine, so it's obvious that if the engine isn't running, there isn't any heat. Don't let it lose power.

We now have pitot heat on to keep the necessary instruments working, and engine heat. Next in importance is the propeller. Jets don't have this problem, of course, and that's a very nice part of the jet world.

✝ The Propeller Is Important

Propellers ice, and when they do, their efficiency drops quickly and then the engine we've kept running is just slinging an icy club around beating up the air but not doing much pulling. Ice can also unbalance a propeller creating enough vibration to shake your teeth.

There are two kinds of propeller deicers: alcohol that's sprayed on the leading edge of the propeller to melt ice, and heat in the leading edges.

There are chemicals available which can be spread on the leading edge of a propeller to help keep ice off. These same products, when rubbed into the surface of the wing deicer boots, make the ice come off more easily, since pieces of ice don't tend to stick to the boot. The preparation doesn't have a very long life and is washed away when the aircraft flies through rain.

There are two ways to use deicing equipment. One is deicing and the other is anti-icing. In the first case one waits until ice has formed and then turns on alcohol or heat to get rid of the ice. Unfortunately the ice never comes off evenly and when it comes off one propeller blade and not the other the unbalance makes a terrific vibration. Also, on a multiengined airplane, the ice slinging off often hits the fuselage with a loud whack that's disturbing to one's nerves. It doesn't hurt the airplane particularly, but any old DC3 will show dents on the side of the fuselage next to the propellers where hunks of ice have beaten on it through the years. Sometimes, too, the ice is tough and will not all clean off properly.

The better method is to use the equipment as an anti-icer. This simply means to get alcohol or heat on before entering icing conditions so as to keep the ice from ever forming.

If there is ice on a propeller and the deicing method used is having trouble getting it off, try running the RPM up and down in surges to give extra, and irregular, centrifugal force to help sling the stuff off.

Propeller ice will often form before visible wing ice, and if one is flying in cloud and an airspeed loss occurs without wing ice, it may be because of propeller ice. I saw this when doing research work with a B-17. We had a stroboscope to look through. It "stopped" the propeller visually while it was spinning at its normal RPM, and you could look at the prop as though it were standing still. A B-17 was convenient for this because you could see the props closely from the navigator's station in the nose. It was amazing to me to see ice on the leading edge of the blades when we didn't have a bit anywhere else on the airplane. And this happened quite often.

So don't be bashful about getting the propeller anti-icing gear in action in advance or under any suspicious condition. The propellers are most important, and if I could have either wing deicers or propeller deicers, I'd take the propellers. If they are doing their work efficiently, you can pull a lot of ice-covered airplane around the sky.

✢ Wing Deicers

We have two major sorts of wing deicers: pulsating boots and hot wings. The boots are always deicers, while hot wings can act as anti-icers as well.

Many times boots are used improperly, and if used improperly they can cause more trouble than no boots at all. The boots pulsate up and down and, in theory, break off the ice. They don't break it off as it forms. The boot has to wait until there is a coating of ice before it can do an efficient job. It is therefore advisable to allow ice to build up at least one-quarter of an inch before turning on the boots. They then expand and break the ice, and the airflow blows it off and away. When they have done this job, it's time to turn them off and wait until another coating of ice has formed and then turn them on and break it off again. Some late model deicer boots have an automatic off-on scheduling device so that the pilot doesn't have to worry about it, but I've never thought much of this gadgetry.

✝ Do Boots Do It?

Actually, I've never thought much of deicer boots. In thousands of hours of weather flying I've never seen a situation where I couldn't have made it without boots, and most of the times I've turned them on have been for amusement.

A good example of this occurred in the B-17. The boots, during those difficult times of World War II, were made of something less than the best rubber, and we continually had problems with them developing tears. Each time we had to replace the boots it was a long maintenance job. Finally, we had a bad tear and couldn't find replacements. I told the crew chief to take them all off and we finished 18 months of weather flying, a lot of it in ice, without boots on the airplane! I was particularly fussy, however, about keeping the propellers clean and the engines running.

Of course the B-17 had a thick, relatively high-lift wing. The modern laminar-flow wings are more sensitive to flow disruption, and boots may be needed. However, care must be taken to be certain the ice is broken off clean by proper boot use technique. A dirty leading edge with ragged pieces of ice sticking to it is often worse than a smooth coating of ice. Boots, at best, are not the most desirable way to deice a wing, and most certainly should not be considered as a means to fly continuously in ice. At the risk of boring the reader I repeat that the worst ice offenders on the airplane are the aerials and gadgets and places where boots cannot be.

I've seen cases where the airplane had a smooth coating of ice when the boots were turned on. Then it broke up in chunks with pieces left on the airplane, even big hunks that stuck on the leading edge and went up and down with the boot, but never blew away. I've seen the airspeed go off ten knots after the boots were turned on and the ice cracked.

Boots raise stalling speed to some degree and must not be on when landing or taking off.

✝ Hot Wings

Hot-wing deicers are something different. They heat the leading edge, and they are best used as anti-icers. When used as deicers, that is, turning them on after ice has formed, they melt the ice; but when it melts, the water often runs back on the wing and refreezes and forms a spoiler on the top of the wing that can materially affect airplane performance; the airplane may develop a severe roll tendency and possible aileron reversal.

Hot wings are more often, and properly, used as anti-icers. They are turned on and the wing heated before getting ice. The hot wing never lets ice form.

☨ You Have to See

An important item that needs deicing, or anti-icing, is the windshield. If we make a low approach through icing conditions and get the windshield covered with ice, we cannot see to land.

The ways to cure this are:

1. Hot windshield.
2. Alcohol to squirt over the windshield.
3. A window one can open to see out.

Historically we have handled the windshield-ice problem in the reverse order. Back in DC2 and early DC3 days we all carried a putty knife in our flight kit. It was used when ice covered the windshield. We'd open the side window, reach out into the icy blast around to the windshield and scrape off enough ice to give a small clear area to peek through while landing.

Then we got alcohol. This would squirt across the windshield and as it melted the ice, the windshield wiper would carry it away or knock it off. It was smeary and partially effective. We still carried putty knives.

An occasional heavy coating of ice wouldn't yield to either putty knife or alcohol, and there have been cases where the pilot bashed out the windshield with a fire extinguisher to have a place to look out!

Now we have heated windshields that seem to do the job well. All the trying days of fire extinguisher, putty knives, and alcohol are taken care of by flipping a switch.

However one does it, keeping the windshield clear is a very important part of ice flying. If the methods available are marginally adequate, then a window that can be opened for viewing is a must! I've always disliked flying any airplane that didn't have a window I could open and at least get enough of a look through to land. A covering of oil from a bad leak can make a windshield no more transparent than a solid wall! The 747s, however, don't have a window one can open—to the best of my knowledge the first transport ever certificated that way—and I flew them about 3,000 hours without a problem due to the excellent windshield deicing system and good wipers.

These are the protective devices against ice. They don't cover the entire airplane. Even if propellers and wings are kept clean, ice will build up in other places, causing drag that eventually will be very serious; which gets us back to the original statement that the first rule of flying ice is to do something to get out of it as soon as it occurs.

✦ How We Fly Ice

Ice flying begins before we ever leave the ground. A number of things need checking. First and most important, is there any frost or ice coating on the wings? If so, we have to get it off before takeoff. The rough surface of the ice can ruin the airflow over the wing, so that the take-off run is very long and liftoff likely to be impossible. We had a Bonanza try a takeoff here at Sugarbush on a lovely fall morning. There was a slight coating of frost, thin enough so the pilot evidently thought it nothing that would affect the aircraft's performance and elected to go. The airplane never got off; it piled up at the end of the runway, and burned—three people dead. Airliners have cracked up because of frost or sticking precipitation on wings. It's a no-win situation and so silly because one is safe, on the ground, and under complete control to get the stuff off before trying to take off. It isn't a matter of being in a situation, rather it's a matter of creating one by impatience. Remember, the most innocent-looking thin coating of frost is dangerous, and flight should not be attempted until it is removed!

Removing frost, frozen rain, snow, or whatever has dirtied up the wing has two aspects: one is de-icing, the other is anti-icing. If it's a nice day and you want to remove the impediments, to use a fancy word, then wash it off with a glycol-water mixture, generally available at the FBO.

If, however, it is precipitating and you are going to take off, then anti-icing is necessary because the glycol-water mixture used for cleaning the wing will only prevent falling precipitation from freezing on the airplane for a very few minutes. So if the taxi is long, and take-off delays in progress, you may be on the ground for considerable time with "stuff" falling and collecting on the wing. Better fluid is required and this is commonly called Type 2, which is glycol-water but thicker, and stays on the wing, giving a certain "holding" time. Coming into use is a fluid called "Ultra" or Type IV, but the iffy factors still prevail. These better fluids have limits, however, depending on what's falling on the wing.

Freezing rain, for example, is the worst, and even Type 2 fluid will only give one to three minutes of "holding" time. It is more effective for sticking snow and other precipitation and may hold for half an hour or more. All this is vague because the time varies with type of precipitation, water content, and air temperature. It's vague enough so that no one has been able to put out an accurate chart to tell how things are going out on the wing as you sit near the end of a runway impatiently waiting for the tower to clear you for take off. Charts list conditions, temperatures, type of precipitation, and hold times, but typically a hold time may be listed as, "From 20 to 45 minutes" for certain snow, or "8 to 15 minutes" for light freezing rain; that's not very precise.

It's one of those things where the responsibility is dumped on the pilot, realistic or not, to be certain the wing is clean before taking off. How are you sure of this? Taking a look at the wing, under strong light if at night, can give a false impression because the ice might be clear and the wing look free of it, when it isn't. Using a flashlight through a cabin window is almost useless as the light glares back in your face. The only sure way is an inspection from the outside, which silly statements and rules say the pilot should do just before takeoff. Imagine climbing out to do that at runway's end, airplanes waiting behind you, wind blowing, wet snow falling, and all the rest. Very unrealistic, but it is your responsibility, so if you are uncertain then the only thing to do is taxi back, inspect the airplane from outside, and de-ice it again if necessary. Unfortunately there's no pat procedure or method, and it's the pilot's responsibility. As the FARs, Federal Aviation Regulations, say in Part 91.3(a), "The pilot in command is directly responsible for, and is the final authority as to, the operation of that aircraft." And that goes be it Piper Cub or Boeing 747.

And an important point: don't think you can blow snow off the wing! If it's super dry and just resting there, perhaps you can, but if it sticks at all, or there's something sticking under the dry snow, you'll never blow it off. This idea is an invitation for trouble. Inspect the wing carefully, see what's under the dry snow, and then if it's only dry snow, get a broom and sweep it off!

If, for some dumb reason, we try to take off with frost on the wing, or other frozen precipitation, we may become airborne, but will probably be unable to climb out of ground effect. The airplane will become very unstable laterally, and it will want to roll side to side, as though falling off on a wing at stall, because that's what is happening. Fly as best you can, you might fly out of it, but chances are you'll never get out of ground effect, which is about half the wing span high. Most times the airplane will not even get that high, but just charge off the runway's end into whatever awaits one there. This is all very dangerous business.

☨ Is Your Airplane Equipped to Fly Ice?

I'm often ask questions such as, "How much ice can a Cessna 172 carry?" How much an airplane can or cannot carry isn't the question. The real question is, "Does FAA say the airplane can be flown in ice?" If it isn't ice approved, then you don't have any business flying it in ice. The FAA approval for an aircraft to fly ice isn't simply a test of how much ice it can stagger around with, but it is an examination of the entire aircraft and its systems. Do the fuel vents stay open, do controls ice up, do air intakes become clogged with ice, can the windshield be de-iced? A host of questions and if your airplane isn't ice approved, you may not simply get in trouble because of the ice on the wing. What if the controls ice and cannot be moved, or a fuel vent clogs with ice? Think about all those things and realize if the airplane isn't ice approved, one hasn't any business flying ice in it. A lot of things could happen.

☨ Propellers and Other Fixtures

The propellers should be clean. On damp, misty, coldish days it's wise to use anti-icing on the propellers right from the start of takeoff. There is some evidence that one can get propeller ice in clear, but very humid, cold air.

There's indication that under these conditions a jet engine inlet may ice, and so engine heat may be wise with a jet engine also under these conditions. Coldish can be anything from roughly −8 degrees C to +8 degrees C.

The airplane's controls should not have any ice obstructing their movement, and the landing gear should be clean, especially if it's retractable. Check the pitot head and static sources closely to be certain they are not blocked by ice or frozen slush thrown up by the wheels during a previous landing, or by precipitation which stuck on the airplane while it was standing.

If it's very cold, be certain there has been heat applied to warm the instruments. A cold gyro may be slow in coming to speed and consequently its action sluggish, which would be bad if one took off and went on instruments quickly. The windshield may fog up during takeoff, and we'd best be prepared to defog it with whatever means are available. After an engine is started it's important to warm it up thoroughly, so that when we take off it will put out and continue to put out its maximum power.

☩ Ice Flying Starts on the Ground

Taxiing can be quite an interesting experience on icy and frozen, snowy runways. First, if there's snow and it has been plowed, it is important to be certain wing tips, and the tail when turning, will clear the snow banks, especially if we are in a low-wing airplane.

The runway surface may be ice-covered and slippery, and a combination of slick surface and strong wind trying to weathercock the airplane makes taxiing an adventure.

First taxi slowly, really pussyfoot along. The most important point is to use the brakes carefully! Use them in little bursts. Don't lock the brakes. As soon as they are held on, the wheel stops and slides, ice-skate fashion, over the surface. Tapping the brake on and off at short intervals will give braking for an instant each time the brake goes on, just before the wheel stops and becomes a sliding ice skate.

I operated my B-17 in the Aleutian Islands of Alaska. The weather there often causes sheet ice to cover the runways and taxiways. The winds blow hard, the taxiways are narrow, and a B-17 with its big fin wants to swing around like a weather vane. I had excellent luck getting around by using the quick on-and-off braking method.

Sometimes on slick runways it's impossible to get the brakes to hold while running up the engine, so one might have to run them up while sliding. This takes thought and planning and careful observation of where one's sliding.

The start of takeoff can be swishy as a crosswind tries to slide the airplane around before there's enough airspeed to make the rudder effective for steering. It takes care, and the winter isn't any time to do things in a rush-rush fashion.

Turning onto a slick runway at high speed for a running takeoff is a very poor practice, because the nose wheel will not have any traction and will only skid as you use it to straighten out the airplane and aim it down the runway. The nosewheel may turn, but the airplane will not, and everything will slide sideways out of control, headed for the boundary lights!

✝ **Where We Find Ice**

Let's look at the places we find ice and how to get out of them. As always we are back to our first look at the general weather picture to see where the fronts are, because in the fronts and low centers we find the most ice and the ice most difficult to avoid.

Ice can be found in large amounts out of fronts, too, but it's easy to avoid if a pilot understands the weather picture. Ignorance can find a pilot desperately in trouble when it isn't necessary.

The classic case of this is the Allegheny Mountains after a low and its associated fronts have gone off the East Coast out to sea. Although the fronts have gone, we find the country from Harrisburg, Pennsylvania, to Columbus, Ohio, cloud-covered. The mountain weather-reporting stations are grim: low ceilings, snow falling, and visibility nil. A pilot flying low, trying to stay VFR, would have a desperate time. It would be impossible to cross the ridges, as they would be in cloud. A pilot flying a few thousand feet higher, on instruments, would find himself getting ice at an alarming rate.

But a clever pilot would know that all this was an air-mass condition with reachable tops above, and it is possible to be sitting in sunshine on top of a seemingly endless blanket of white, relaxed and comfortable.

Countrywide, such air-mass stratocumulus decks are common in winter after fronts pass. The new cold air is unstable and builds a cloud deck. On the ground, after a front has passed, we are often under clear skies. But cumulus start to build. At first there are pretty white scattered CUs; they become broken and dark and finally turn into a gray overcast sky that spits snow in blustery cold winds. This is the real cold air mass moving in.

That cloud layer has ice in it. The only decent place to be is on top. The tops will vary in height depending on a number of things.

Tops are higher in mountainous areas or to the lee of large bodies of water, such as the Great Lakes, where the air picks up moisture. The tops will be higher closer to the frontal system, though not directly behind it, because there is often that clear area where the new unstable air hasn't had enough time to get in and start its action. This clear area isn't very wide. It often fools people into thinking everything is wonderful ahead, that the front has passed and now it's clear. These hopes fade as clouds begin to form and bases get lower and tops higher 100 miles or less along the way.

As we get farther and farther from the weather system, the tops are lower and the bases higher, and the showers less frequent until they die out altogether. This happens because the air has been in the area long enough to be modified, and its instability is reduced as the air temperature and ground temperature become more nearly the same.

✢ Temperature Again

This temperature difference between ground and air is the reason why we often see beautiful, clear, cold, winter nights and, as we look at the sparkling stars, decide tomorrow will be a lovely day. Tomorrow turns into an overcast, cold, gray day. Why? Simply because when the sun came up it warmed the ground; maybe it didn't feel warm, but it was warmer than the air, and this started the air rising and triggered the same process that makes cumulus clouds in summer.

✢ Where Are the Tops?

If we are going to fly in the area of the heavy deck with its ice and snow, we should try to learn what the tops are before we take off and start climbing up to find them. Ask the FSS, Flight Watch, or the Weather Service. If they don't know, ask an airline pilot who has just come in, or call the operations department of an airline and ask them to ask a pilot who's taken off or landed. The airline pilot is happy to help out. And when you get up there, don't keep it a secret. Call the nearest Flight Watch or FSS and tell them what ice you found climbing up through and what the tops were.

If we are climbing to top a deck, it's best to climb quickly—not at a nose-high staggering angle so that ice will form back under the wing and hurt, but at high airspeed and with plenty of power. Stratocu clouds have the most ice near the tops, so don't struggle along clipping through the tops; get up and out of the clouds.

Occasionally there are bits of trickery in getting on top. Out of Pittsburgh, Pennsylvania, headed east, the highest tops and toughest ice are located in the mountains east of Pittsburgh, over Laurel and Chestnut Ridges. Taking off and then climbing toward the east means one climbs up through the thickest clouds. To the west of Pittsburgh, of course, the land is lower. Back in DC2 and DC3 days we often got a clearance to take off and climb toward the west. The tops would always be lower that way. Once well on top we'd turn back east for New York. It was a way to get up through the minimum amount of cloud. The tops to the west would often be 7,000 feet while the tops over the ridges might be 12,000 feet.

Flying in winter over the mountains of the Far West can be difficult, with a combination of moist Pacific air being lifted up the mountains of the coastal ranges, the Cascades and Sierras. It can be a problem, at various times, from the Canadian border to Mexico. The fact that the airplane must operate at extreme altitudes because of the high mountains—higher than those of the eastern states—adds a trying factor. In fact, with certain lower-performance aircraft, it is an impossible one. While the West Coast is generally blessed with good weather, when it is bad, it is *very* bad! Be certain to check what the tops are before takeoff. If you're planning to fly on top, don't experiment and find yourself still trying to climb with much ice on the aircraft and no tops in sight!

Descending through icy decks should be done quickly without being wild and panicky about it. Ice raises the landing speed, covers windshields, and makes a low approach tougher. We don't want to get any more ice than necessary during the descent.

ATC will often hold or dawdle you about in the icy deck, and when they do, it's time to tell them that you are getting too much ice and you don't want to sit in the deck any longer than you have to. They'll generally help if you let them know you have a problem. It's important, then, to know if cloud decks are air-mass or frontal, and what the tops are.

✢ Fronts and Ice

Knowing the front locations is the most important factor in foreseeing ice, but there are other hints, too. If a cloud deck exists and the surface reports show a good, solid wind of at least 10 knots (not a light, variable wind) from a direction we normally associate with good weather (like NW in the East Coast area), we can feel pretty certain that the clouds are air-mass and not frontal.

If we are on top, it should be clear above us, or perhaps we may have only a very high, thin cirrus deck left over from the system to the east. If, however, thicker high clouds are overhead and we are between layers, we'd better check to be certain that something isn't stirring in the general weather picture.

The first thing to do is look westward, or in the direction away from the weather movement. If these high clouds thin and decrease in that direction, with perhaps blue sky peeking through and a general clearing "feeling," then the clouds above us are probably leftovers from the past weather system.

If, however, it looks just as cloudy to the west, we must be suspicious of another weather system moving in, or the one to the east being stalled, or perhaps an occlusion bending back.

With this gloomy outlook it's time to check surface reports in all directions and try to get any new or revised forecasts. Once I carried blank maps in my flight kit and I would draw up a crude weather map in flight from weather reports received in the air. I could quickly see a change in wind circulation and begin to notice a change in weather patterns. This proved to be useful in a number of ways, like seeing a trend for deteriorating weather or for better weather; and it was excellent training to develop a deeper awareness of what weather does.

Air-mass clouds may have snow showers, but they don't have large areas of steady snow. Steady snow or rain indicates that something more extensive than air-mass weather is in progress. This we can see from surface-weather reports. An exception is in mountains with unstable air being lifted. Then the snow might be steady with very low ceilings on the upwind side, but just snow showers elsewhere. The steady snow is the result of the wind constantly lifting the air up the mountain. The same effect can be found in the lee of lakes, especially the Great Lakes. Even Lake Champlain in Vermont and New York can cause heavy snow downwind from it.

✝ Equipment

When we fly fronts that have ice in them, we need the proper equipment. This time the equipment is the airplane itself. It should have the ability to climb to a fairly high level without much trouble, 15,000 feet for example, and it should have a good enough cruising range to give a wide choice of alternate airports. .

✝ Not Always in Clouds on Instruments

We have to define something before we can go much further in ice flying. That something is the difference between being on instruments in a cloud and being on instruments but not actually in solid clouds. Sounds a little complicated. Say you are flying in dense haze or smoke with a lower cloud deck that obscures the earth. In this condition you are not really flying in a cloud, but you can't see anything, and so you have to fly by instruments. Now take this condition and make the haze snow, and you have the same thing: You are not really in a cloud, yet you are on instruments. This occurs very often in winter instrument flying. It's not a question of cloud density; it's just that you are either below a cloud deck or between layers of clouds in snow or haze. Whenever this occurs you do not get ice! The only time you get ice is in an actual cloud (except for freezing rain). You can fly at times for hours in such cloudless instrument conditions and not get any ice even though it's below freezing.

✝ **Warm Front**

Taking it from the ground up, as if you were taking off and climbing out of a field, the structure of a warm front, and any condition that involves warm air overrunning colder air, is something like this. At first there probably will be a low cloud deck, its base at perhaps 500 to 1,000 feet. This deck is caused by the precipitation falling into the colder air, and its thickness will depend on how long the precipitation has been falling and how close the frontal surface actually is. It may be scattered and only 1,000 feet thick, or it may be solid and 8,000 feet thick. It will have ice in it, probably light rime ice, although it can be moderate at times. As you climb through this layer it will be snowing and the windshield will get ice on it and so will the wings and other parts of the airplane, and you'll need alcohol and all the rest. The wing tips will appear fuzzy and you can see the shredded cloud slipping over the wing. At night the running lights will be a fuzzy glow. Then suddenly you notice that the fuzziness on the wing tip has gone and the running light is bright and sharp, and if you look closely you'll notice that ice has stopped forming. You are still on instruments and in snow, but you have gotten on top of the lower deck. Now you are between layers. Staying here you will not get any more ice, just snow, which doesn't bother you because you are not actually in a cloud. The air will generally be smooth.

Occasionally you may feel a slight bump, and if you look out, you may see a slight fuzziness on the wings and realize you have picked up a little ice. What happened was that you went through a portion of cloud. In doing so you probably encountered temperatures that were fairly close to the freezing point. This little cloud is just a piece floating around in the mess trying to form, but it hasn't got quite the right formula, and the water vapor would rather go directly into snow. The lower the air temperature the truer this will be; that is, the colder the air, the less cloud. Thus we can form one rule that says, "If the temperatures are near freezing, warm fronts are dangerous things and should be considered very carefully before being flown."

At higher levels in this vertical cross-section we may find another cloud deck; but it will be high, 14,000 feet or more, generally composed of ice crystals, and have only light rime ice. If this deck has more than light ice you are probably in the slope of the frontal surface. Then go down to get back in the unclouded snow, or if the airplane climbs well, go up to where it's too cold for much ice. This will be above 18,000 feet, perhaps 25,000 feet.

It's wise, also, to keep the surface and lower-level temperatures in mind and know what they are. Sometimes it's easy to descend into a region of above-freezing temperatures and, of course, no ice. It's pretty silly to sit up high fighting ice when one could be lower in above-freezing temperatures.

Sometimes this other cloud deck actually doesn't exist. I remember one time flying in moderate snow at 8,000 feet. I was on instruments but not actually in a cloud. For curiosity's sake I climbed on top. The top was 25,000 feet, and at that altitude there was only blue sky above. During the entire climb the airplane never went through any actual cloud, but once I was on top looking down, the stuff below seemed to be definite cloud and it looked like the top of a stratocu deck. Upon letting down into it again I found only snow and haze, no cloud and no ice!

Now as the frontal surface approaches, cloud becomes more frequent. If there is a top deck, it will eventually merge with the lower deck, and there will be a fairly thick cloud deck. Here is where the front becomes interesting, because here is the ice problem in that wall of cloud. The lower the temperature, the less ice; but just for meanness' sake Nature fixed it so that you can get ice at pretty low temperatures. It's the moisture content that really counts, and that you'll have to have gotten from the meteorologist before you took off. The meteorologist can look at the radiosonde observations and get a pretty good idea of how much moisture there is aloft. Of course, as we've said, the chances are slim that we can talk to a meteorologist; so what then? It will be in the forecast which will simply say, "Icing in clouds." If there's lots of moisture and conditions are right for ice, the forecast may call for severe icing. That's their way of saying how much moisture there is. It doesn't give us a "feel" for it, and they may include a fanny-saving fudge factor to be conservative, but we cannot count on that. Anyway, it's on the safe side, and if the forecast says ice, it will probably be there.

Generally, the distance through this walled area isn't very great, but it can be far enough to give a load of ice that's very troublesome. The easy way is to go on top of all of it, climbing clear on up until it's CAVU above. Unfortunately, the tops of some of these conditions are very high, in the 30,000s, and you need a high-performance airplane to do that. The more laborious way is to barge on and see what happens, but first the pilot should be pretty certain not to be flying along the front but through it via the shortest distance. This information can be obtained from the weather map before takeoff by studying the location of the fronts and the probable direction you'll go through.

✛ Fishing to Get Out of Ice

In flying up to the front you've stayed at a level that is pretty much cloud free. That comes from old Rule Number One: Do something about ice when it first forms. If you are flying in snow without ice and then ice suddenly forms, you start a fishing expedition by climbing if low and perhaps descending if high, carefully watching the wings during the climb or descent until you're out of cloud again, noted by the lack of fuzziness. This climbing and descending will depend on the temperatures. If it's very cold it will only take a little altitude change to get out of the cloud, perhaps as little as 500 feet. As the temperature increases that cloud will get thicker and perhaps run into thousands of feet.

When we learned these things, and experimented with them, it was the era of DC2s and DC3s. There was very little ATC anywhere, so one could change altitude and fish up and down, sometimes only a few hundred feet or so, without asking for or getting a clearance, a happy time. Today, of course, we need ATC clearances before changing altitude, unless it's an emergency, and then we must announce it as soon as possible.

Now as the frontal surface is reached, climbing or descending will be a waste of time because things are all pretty much cloud, unless we know that there are above-freezing temperatures below. The pilot ought to know where the frontal surface is apt to be to avoid climbing or descending when it is useless to do so, and by the same token to not just sit at one altitude in ice-producing cloud before getting to the front. All this means a careful study of the weather map and weather reports before takeoff, so that the pilot can decide where the front is, what its past movement has been, and where it will probably be out there.

Another way of telling when the front is reached is by the pre-cipitation: It will probably increase considerably and also the clouds will be more turbulent. If the precipitation becomes heavier, the air rougher, and the cloud thicker, you have reached the front.

Now it's a question of going through and hoping it isn't too far. The best out now (an out is something no sane airman takes off with-out having well thought out) is a 180-degree turn. You can poke on in there and start getting ice. You know what's behind you and know where you can fly to keep out of ice. Now it's a question of deciding how much ice you are willing to take on before you turn around. This amount must be divided in half, because if you go in and then decide to turn around, you'll have to go back through that much. Most times it isn't too far through, but it's something that makes you sit pretty much on the edge of the seat. Once through, there will be an abrupt change of some sort, the turbulence will stop, the precipitation will let up, or you may break out between very definite layers or on top. Once a pilot feels the front has been flown through, it's time to fish again for a top, a layer, or something ice-free.

✝ Taking Off in a Front

It can be dangerous to take off when a front is very close to the station. You are apt to climb right into a mess that you don't know much about. It's best to wait until it's passed and then sneak up on it knowing a little of the stuff it's made.

✝ Learning Time

All this talk has been about fronts that have below-freezing tempera-tures throughout. In spring and fall, however, fronts have above-freezing temperatures in various places, and while this sometimes helps, sometimes it makes matters worse.

If there are above-freezing temperatures in lower levels, it's a good time to play a little. You know then that your out is to descend and melt off the ice. It's a good day to fly high and find out what the fronts are made of. Before takeoff a close study of the weather maps and reports will give you a good picture of what is happen-ing and where the above-freezing conditions are. Then you can go play.

But above-freezing temperatures can also be dangerous. Suppose the overrunning air is above freezing and it's overrunning below-freezing air. The precipitation will be rain, except in the lower levels. Then it will be freezing rain, and that's nasty stuff.

✝ Orographic Effect Again

The thing that complicates all the front business is local effect orographic lifting. With the extra help of air being lifted up mountain slopes by the wind, the air can hold lots more moisture. When air is lifted that way you can get severe icing at very low temperatures. Take a cold front over the Allegheny or Cascade mountains. It collects lots of extra moisture that is pushed up. That means more ice.

These orographic effects are peculiar to various regions and unless you know the region they must be guarded against. Here is where the meteorologist comes in, and a pilot must always ask a question about orographic effect. With no meteorologist you inspect the prospects on your own. One, are there mountains to cross? Two, is the wind flowing up the mountainous terrain? Three, is the wind circulation coming from a source of moisture like the Pacific Ocean or Great Lakes? If all the answers are yes, you can be certain there will be lots of ice in the clouds, and you'd better be prepared to cope with it, or sit on the ground. Always be extra cautious with ice and fronts in mountains.

✝ Cold Front

In flying fronts the problems are pretty much the same whether the front is cold, warm, or occluded. A cold front is more violent, quicker to get through, rougher, and has more ice. The warm and occluded fronts are a little slower but cover a larger area and make you sweat longer.

Ahead of a cold front there is a high deck, with some layers below, but it's generally easy to maintain a position out of the clouds and of course out of the ice.

As we fly into the front, ice will be plentiful. Cold fronts are unstable, with air being lifted quickly and therefore carrying considerable moisture. The ice can accumulate quickly and heavily on the aircraft.

The distance through this area is relatively short, however. The danger, or rather the mistake often made in passing through a winter cold front, is in staying in the stratocu deck that forms behind it. The front itself may have been passed through, but the airplane remains on instruments getting ice. Actually a couple of thousand feet higher might well put the airplane on top, in the clear.

Cold fronts can be flown higher than warm fronts to avoid ice. The lower the temperature the less ice, and so flying up high will help. Then, when the actual solid part of the front has been traversed, there's a better chance that the airplane will be on top of the lower clouds on the backside of the front.

But flying weather of this type is like learning to play the piano. It takes time. A pilot must study the weather carefully and then begin crawling before walking. There are certain days when we can poke into these things and see what they are made of and still leave an out or two. For example, when overrunning air causes an altostratus deck, the weather is simply high overcast, with the weather at the airport good. That's the day to go on up there and poke into that altostratus and fly a while toward the frontal surface, then turn around and come on out and mull it over. When any opportunity like that presents itself, it should be used.

We've flown through ice, made a letdown, and now have the runway in view. The flight is about over and we've got it made.

We really haven't, of course, because that runway may be ice-covered. Now what? First, we land short to have as much runway as possible for braking, but not so short that we undershoot. Once on the ground we want to try the braking gingerly and see what we have. We want to do it early in the rollout. The best way is the on-and-off method we talked about before. Airline and big corporate aircraft are equipped with antiskid brakes, which does all this off-on business automatically. Now automobiles have anti-lock brakes, which, to a degree, are the same thing. The modern sophisticated airliners have automatic braking that goes into action on touchdown and brings the airplane to a stop without the pilot doing a thing.

There's always a question about how far to go with aerodynamic braking; that is, using flaps during the landing roll, or trying to keep the nosewheel up in the air and the tail down so that the entire airplane is creating drag and slowing down.

It's a known, established fact, of course, that the fastest braking comes from wheels in contact with the ground. This means that one should get the nosewheel on the ground gently, the flaps up so that lift is lost, and the airplane's weight on the wheels where the brakes can take over. This is the best way to stop and remember the on-off-on-off brake technique.

But if the runway is sheet ice, there may not be any brakes at all until one is moving quite slowly, and braking will be marginal then.

So it would seem that aerodynamic braking would be best; that is, we should get as much as possible from it. But if a runway is that slick, then we really don't have any business landing on it. We should go elsewhere if possible.

The facts say to get the weight on the wheels and get stopped, but there may be a few times when this won't work. We cannot make a rule. This is where judgment backed with knowledge takes over. Pilots, I hope, will always have to make judgments; it wouldn't be a very interesting game if they didn't.

The luxury of reversers, jet or propeller, makes a safe and graceful stop more likely. In the case of a jet remember the reversers are most effective at higher speeds, so get them in action as soon as possible after touchdown and don't be bashful with the thrust. Be careful to remember that in crosswinds one may get going sideways and headed off the runway with reversers, and we have to be alert to that possibility.

But let's say we are well slowed down on the runway and feeling pretty good about it. Now another little hazard creeps in: turning off the runway. It's often tempting to turn off as soon as possible if there is landing traffic behind us that we don't want to hold up.

But too fast a turnoff on a slick surface may find us going sideways into a snow bank. One must proceed slowly on icy surfaces. Generally speaking, you are rolling along at a faster rate than you really think. When feeling you are doing 30 miles per hour, you're probably doing 50. Turn off the runway onto a taxi strip slowly, even if the tower is yelling at you to clear the runway; if runways are slick, then ATC had better give a little extra spacing between landing aircraft.

And as we park the airplane, it's important to be certain it's not on a slick surface that will allow it to move after we've left it. Surprisingly, too, pilots sometimes hurt themselves when getting out of the airplane and setting foot on an icy surface. Our minds are apt to be so full of reflecting on the flight and landing just made that we don't get mentally back to earth soon enough to think about such details as ice on the ground. An undignified fall with a broken bone or two would be too bad, especially after flying safely through all that weather and ice!

16

Taking Off in Bad Weather

Now let's talk about the three parts of a weather flight: takeoff, en route, and landing. We have studied the weather in earlier chapters. We've made a flight plan, picked a surefire alternate, and decided on fuel. The airplane has been preflighted. It's ice-free, and the pitot static ports are free of any obstructions.

The charts we need are arranged in order and placed where we can get them easily in flight. We've studied them to see what the routings may be, including departure from the terminal, en route, and arrival at the destination.

There's a pad and paper handy—perhaps a knee pad. We are wearing a boom mike if we're lucky enough to have one. We do our normal checklist and check each radio and set it for the stations required for departure. If there's a local broadcast ATIS (Automatic Terminal Information Service) that tells which runway is being used and departure information that includes altimeter setting, we've listened to it.

✛ Altimeter Setting

If there isn't an ATIS, we get the wind and altimeter setting from the tower or FSS. Altimeter setting is very important and should be set before takeoff. Don't just set the altimeter to the field elevation and think that's okay, but obtain and use the altimeter setting. This checks the altimeter's accuracy by noting if it reads the field elevation when set. If it doesn't, then our altimeter's accuracy is suspect, and we'd better dig deeper. First, get a repeat of the setting, and if that doesn't correct the problem it's time to go back to the shop.

If there's an omni check on the field, the ship's receivers should be tested for accuracy. When checking the ADF, be certain it's checked in all its functions: antenna, BFO, manual loop, and pointing to the station when in automatic.

✝ Be Prepared

Unless there's a directional beacon that is part of the departure procedure, and this is rarely so, tune the ADF to the outer locator of the ILS runway being used for landing. This is an emergency item in case something serious happens right after takeoff, such as a fire, and you want back in fast. You'd know which way to head as you tuned the ILS and told the tower that you were in an emergency and needed an immediate return, and what you were doing. Unfortunately the FAA is removing the NDBs (nondirectional beacons) from outer marker locations; this, in my humble estimation, is dumb.

It's a far-fetched possibility, but why not be prepared? This also makes it important to check which runway is being used for landing, what the ILS frequency is for it, and the inbound heading. The inbound ATIS can be checked for this, or you can listen to tower arrivals.

✝ Let's Go

Now we start up and call the tower for taxi and airways clearance. As we taxi out we get the pitot heat on. If it's a warm day, put it on just before takeoff. During taxi, check to see that the directional gyro and turn indicator are working and indicating turns in the proper direction.

When the tower calls with the ATC clearance, be ready to copy it on a piece of paper—the knee pad again. Never leave it to memory. After it's copied, read it back to the tower.

After that's done, let's absorb the clearance. First mentally picture the route. Be certain the radios are set to the stations and radials required. Then firmly implant the altitudes in mind.

Most clearances request changing to some frequency for departure control after takeoff. Make a firm mental note of this frequency. Write it as part of the clearance. Also, once in the air, when new communications frequencies are given, write them down. In this way you'll always have the last frequency. If, then, you call on the new frequency and there isn't any response, you'll know what frequency to go back to. There's nothing more frustrating than to tune a frequency, discover it's a dud, and then ask yourself, "What the devil was that frequency I just left?"

But let's not be hurried about changing to departure control. There's a tendency for pilots to feel they must change as soon as the wheels leave the ground. Well, the most important thing right then isn't departure control; it's flying the airplane! Before changing frequencies and talking on the radio, let's get squared away, climbing, and with things set properly. A copilot, of course, makes all this easier and faster. But always, as in everything we do, the first thing is to fly the airplane.

Good use of radios will make frequency changes easy if we plan in advance. Since most airplanes have at least two communications radios, one can be set on the tower and the other on the departure control frequency. After takeoff, when they say to change to departure, simply flipping the control panel switch to the other radio gets it done with a minimum of distraction. The newest radios have fancy storage capability, and one can have two or more frequencies all set and call the one wanted into use by flipping a switch or pushing a button. These are useful in cruise, too; when a new ATC frequency is called for, the old one is still stored and can be recalled in case the new frequency doesn't seem to raise anyone.

As I said, mentally go over the route you'll follow after takeoff. This means to visualize the takeoff and departure, then make a quick review of the basic stuff like, "Lift off, up gear, climb at_____indicated. Tell tower I'm changing to departure control. Up flaps at_____, set power. Right turn to one five zero. Level off at three thousand," and so on.

The point is to mentally preplan, to know in advance. Don't just fling yourself into the air and then begin to think what to do. It's all part of the important rule about flying well ahead of the airplane.

✛ Don't Be Bashful!

Of course, the tower departure control or ATC may suddenly change it all, and you have to be quick and flexible. Again, write it down, visualize it, and if there is the slightest doubt, ask ATC for clarification. Don't be bashful! The neat cockpit with handy charts is also important here. It makes it that much easier to look up some intersection or omni that you've never heard of. There isn't anything wrong, either, with asking ATC where the heck the intersection or station is, if you can't find it.

When tuning or retuning radios, be certain to identify the station and have the identification sink in. There have been embarrassing things occur because of failure to identify, such as a pilot I knew who tuned Orly, Paris, France ILS, shot an approach to 300 feet, and landed only to find he was at Le Bourget! He failed to identify the ILS. But things have happened with more serious consequences, so identification of any aid before use is very important!

It is very easy to become complacent about this because the numbers on radio dials are clear and there isn't any fishing and tuning; simply turn the dial to the number, and that should be the station. But mistakes can be made: you looked at the charts for a frequency and set it up, but in the chart confusion you read the frequency for a different station; frequencies are changed and we may have missed the NOTAM, or we may not have the latest chart revision in hand. All these, and more, have caused wrong tuning and accidents. Even our home station, which we know so well and, perhaps, are looking at right over the nose, should be identified. Identification must be a habit!

Altitudes are also tremendously important. The ones cleared to should be written down or placed on an altitude reminder. Altitude reminders on airline and many corporate aircraft have warning sounds that alert that the altitude is being approached or that you are wandering off a selected altitude. These are good gadgets, but the pilot still should keep altitudes firmly in mind and not count on the gadget alone to do the job. Altitude reminders have numbers you can set mounted on the control wheel hub or instrument panel. It's a substitute for writing them down—a good one, too.

Most all aircraft have Mode C transponders that read altitude so ATC can catch any diversion, but as in all flying, we should be certain to pay attention to altitudes required and not count on the Mode C alone. Your checking altitude and the Mode C broadcasting gives the redundancy desired in all flying, the important double-check.

One should anticipate altitudes. If cleared down to 6,000 feet, one ought to be thinking, at 7,000 feet, "I'm leaving seven; remember, level off at six." In other words, climbing or descending, be at least 1,000 feet ahead of the airplane. If it's a fast jet descent, be 5,000 feet ahead.

When ATC clears us to a new altitude, we should get right at it and leave the present altitude as soon as possible, unless, of course, ATC says we can leave at our own discretion. I get a creepy feeling that when ATC says to leave a certain altitude, the pilot fiddles along for some time, or makes a little halfhearted effort at some minuscule rate that vacillates between 100 fpm and 300 fpm. GET at it! I get that creepy feeling because I wonder if the person they've cleared into my old altitude has started toward it promptly and perhaps too fast and will get there before I leave! When cleared, one should descend or climb smoothly but promptly. If for some reason such as waiting to reduce speed, one cannot, ATC should be informed.

✝ Off We Go

The tower clears us for takeoff and we swing into position. If we are departing into low ceilings, we hesitate a moment and let the gyros settle on the runway heading. Check the compass, too, and see if it agrees with the runway heading. (Check landing, too, and be certain you're headed for the correct runway!)

If the time from startup to takeoff is short, delay the takeoff, in the warmup area, for five to ten minutes to be sure all gyros are up to speed. If it's night, be certain the cockpit lights are set to the best level for you, and that your eyes have accommodated to that level.

As we start the takeoff, we make a last-minute check of these few items, what I call the "killer" items: carburetor heat off, fuel on fullest tank, flaps where they should be, propeller in low pitch, pitot heat on. The regular checklist should have prepared us well, but it doesn't hurt to have a few, most important, items that can be checked almost in one glance. My last check in a 747 takes a second or two as I look at fuel quantity, pumps on, flaps at takeoff, stabilizer set, spoilers down, heading in agreement with runway, and this is after I've completed a checklist. It's a habit that ensures that things that could cause trouble on takeoff are set okay.

✝ In the Stuff Quick

We should have in mind that once we lift the nose and leave the ground we'll be on instruments. It's no time to worry about flying VFR first and then transitioning to instruments; be ready at once to be on solid instruments. If there's good ceiling and visibility then, of course, we want to keep an eye out for traffic, but let's suppose the ceiling is low, 200 feet or less. Also, most night departures, low ceiling or not, require reference to instruments more than to an outside horizon that's poorly defined, if at all.

Going on instruments means going to attitude flying immediately; it means looking at the artificial horizon. With a certain amount of nose-up on the bar, at takeoff or climb power, we must be going up. One should know the pitch attitude for one's aircraft for climb out, and then fly it on the artificial horizon. This is also useful if, for some reason, airspeed indication is lost or obviously erroneous. So that's the first step: fly attitude.

If the wings are level we are going straight, and if one is down we are turning, unless we're handling the rudder like a new student, which we shouldn't be. So the position of the wings on the artificial horizon is the other thing we look at.

With wings level and the nose where it produces climb, we are on a straight course and climbing. It's really very simple. If we have to turn, we lower a wing: 30 degrees is enough. ATC doesn't expect more, and making a steeper turn means more up-elevator and more concentration that may divert you from other tasks. Keep it simple.

We first set the attitudes we want and then cross-check the airspeed, heading, vertical speed, and altimeter. Close to the ground the pressure instruments briefly give erroneous readings. Watch the vertical speed and altimeter just as you take off some clear day. Chances are the vertical speed will show a descent and the altimeter will start down until you are 50 feet or so above the ground.

Once away from the ground the pressure instruments settle down, but even then attitude is the primary guidance and the other instruments a cross-check. This cross-check is frequent, however, as a part of our constant scanning process.

✛ How about the Weather?

We are off and away, but so far we've talked about flying the airplane; so let's go back now and talk about the weather part of the takeoff.

Before takeoff, if we think it's cold enough for ice, we should have the propellers slicked up with alcohol or hot with heat. The pitot heat is on, as I've said. If we are flying a jet, engine anti-ice is on. With piston engines we'll clean out the carburetors with heat just before takeoff and then remove the heat as we start to roll. But we must watch for engine icing conditions once in the air.

Before takeoff, check the wind direction and velocity. We learn any crosswind we may have to compensate for. The velocity says not only how fast we'll get in the air, but if it's gusty, it tells us if we can expect turbulence when in the air. And a thought about the wind just above the ground will remind us of any shear possibility.

The runway surface is of interest. If it's winter and the runway icy, our ability to stop if the takeoff is aborted will be materially reduced. The V_1 figure is out the window if the runway's icy. It isn't much better if the runway is wet. The slick surface will make stopping difficult.

The combination of a strongish crosswind and icy runway calls for care in taking off. Until enough airspeed is obtained to give rudder response, we're dependent on nosewheel or tailwheel steering. Neither has much traction on ice, and so the beginning of the takeoff is a little dicey, as the English say.

There's always an anxious moment, but this can be improved by getting lined up on the runway before starting the takeoff. Swinging onto an icy runway on a running takeoff is an almost sure way to be partially out of control right from the start. Getting lined up, stopped, and all squared away before starting to roll is the way to do it. Don't let the tower push you into a quick takeoff just because they are trying to move traffic faster.

Snow and slush will increase the takeoff run dramatically and should be taken into account. Airlines don't take off if slush is over one-half-inch deep. Standing water in deep puddles on the runway will prolong the takeoff run, though not as much as slush.

Any takeoff, even in very reduced visibility, is started visually. We need something to grasp with our eye. If the visibility is one quarter mile or less, we need a good centerline stripe or, better, centerline lights. Without these, the visibility must be sufficient to see the runway far enough ahead for guidance.

Once in the Air

Keep the runway heading well in mind, and once in the air, fly that heading until a turn is required. Both landing and taking off in snow, we find the air choppy and turbulent for approximately the first 1,000 feet. This is almost always true in snow because of an unstable condition caused by the release of heat during nature's snowmaking process. It isn't dangerous turbulence, but it makes an instrument approach or a precise departure procedure more difficult.

Thunderstorms Again

In summer, with thunderstorms in the area, we think in other ways. Thunderstorms mean turbulence and the possibility of sudden wind shifts. They mean heavy precipitation, downdrafts, and shear. All this during takeoff, at low speeds and close to the ground, makes us quite vulnerable. If storms are very close, it's best to sit on the ground until we have some maneuvering room to get in good flying condition and avoid them.

With storms in the area when we take off, these are the things to think about: Have radar equipment warmed up and ready for action. With a spherics system on board, one can inspect the entire area around for thunderstorms before takeoff. Most radar equipment has a "standby" position in which the equipment is warming up but isn't actually putting out signals until the control is flipped to an "on" position, when the antenna begins to search and the tube lights up. We don't want it on in the ramp area, where fueling equipment, people, or other airplanes are nearby. A pilot needs help using radar right after takeoff because attention must be centered on the business of takeoff, not on a radar scope trying to interpret the picture. Takeoff and early climb out demand a pilot's primary care.

Someone else is needed to study the scope and tell the pilot where cells are and how much turn is needed to miss them. Pilots, of course, can reverse these duties, but they must be organized so the proper attention is placed on each aspect of the flight.

The flight path one would like to fly after takeoff can be worked out with ATC before takeoff if inspection by spherics or radar has shown where thunderstorms are. It's so much more relaxed to do these things on the ground than in the air, while on instruments and busy.

While we are trying to duck thunderstorm cells, ATC has to be kept advised and give us advance permission for turns if possible. This is especially true in a terminal area.

We have to think about the wind during takeoff and how close the storms might be: It could be possible to get a hefty, gusty wind shift with shear and downbursts during takeoff.

Suppose we are taking off southward into a strongish wind. A thunderstorm arrives and snaps the wind to northwest; our takeoff run is lengthened, and once in the air, if we get there, we'll probably be in downdrafts that will make the climb sluggish at best and more likely dangerous! It's a silly time to be taking off, of course, but sometimes it is difficult to tell just where the storms are if lower clouds hide the actual storm clouds.

But we shouldn't be taking off, or landing, with a thunderstorm any closer to the airport than our rules for normal avoidance in the air call for: 10 miles minimum and 20 is better. The airplane is in poor condition, accelerating from slow speed during takeoff and at low speed during initial climb; a strong shear during this regime of flight is indeed dangerous and difficult or impossible to cope with.

Once in the air we may get some pretty wild turbulence. What we want is to get up to a good workable airspeed as soon as possible; but, of course, we don't want to shove the nose down and fly back into the ground either. It all takes care and judgment.

We should have in mind a direction to run away from the storms if things get too tough. This knowledge comes from a careful study of the weather before takeoff in order to know how the weather is moving. Study the sequence reports for stations in the area, a 100-mile circle at least, to see what is actually happening. This sort of weather is tough, and it shouldn't be taken lightly.

⌁ Thinking

Now we are in the air on instruments, encased in the tiny world of our airplane, everything outside ending at the windshield. Let's not worry about relating ourselves to the outside, trying to see out there or holding on to that world we left. It's best to forget it, settle down, and watch the instruments.

We concentrate on the rather simple job of flying instruments. It is more simple the more we are relaxed. Lack of worry makes us relaxed. Uncertainty makes us worry, so let's not be uncertain; let's know what we are going to do next, be prepared, and plan ahead.

Our thinking is in layers. The top layer is flying instruments. But this is easy and doesn't require much of our thought. If we've practiced enough, it's almost automatic.

The second layer is the airplane. Are we doing the right things to keep it running? Carburetor or engine heat? Power settings, pitot heat on? Fuel on proper tank, the gear and flaps cleaned up? In other words, all the things that will ensure the airplane's running. If it stops running, so does the flight, so it has to run.

The third layer is navigation. What's the radial we are trying to follow? Is the station identified? What's the altitude? Are we level or approaching it? That intersection coming up, are we tuned, identified, and set up for it? We review all the items that are a part of our navigation, including flying that heading the person on the ground may have given us.

Layer four is the weather. Turbulent? Fly turbulence methods. Ice? Is all the anti-icing or deicing equipment being used, and what does the clearance ahead do for our getting out of the ice? Thunderstorms? Are we ready for turbulence, is the radar or spherics working, and do we know a way out if we want to run?

Layer five is a quick thought toward emergencies. If I want back in, how will I do it? If I cannot get back, where's the nearest place I can get in?

These layers are not precise, nor are all the items listed, but the thoughts for building your own are there. The layers intermix at times, but once we get the airplane and engines set, we don't have to think about each thing constantly. If the weather equipment is on and working, we don't have to think about it constantly. The emergency action, once it's implanted in our mind, doesn't require attention; it's there if called upon. Each layer has a priority based on how often we need to refer back to it.

As we navigate or follow radar vectors, it's wise to visualize where we are and what's under us. We shouldn't concentrate on this to the extent of neglecting anything else, but we should keep in mind roughly where we are and what we are over. This is particularly important in connection with terrain height. How high is the terrain and are we clearing it with enough margin?

The charts give minimum altitudes, and we should pay attention to them. This is a part of preplanning; before takeoff we should have a good idea of where things stick up, be they TV towers or mountains. All the fancy equipment and radar vectors should keep us above the terrain, but in the final analysis, missing the ground is our responsibility. In the back of our mind should be an awareness of what a topographic map of the region looks like, especially the heights of terrain, and then what our actual altitude is; this, as they now like to call it, is situational awareness. Before modern equipment and ATC to lead us around, we always flew with map in hand, location and height known and related to that map. Unfortunately, in today's world, maps—topographic maps with mountains, rivers, lakes and all the rest—aren't used very much. The radio facility chart has become the paper that pilots hold, but radio facility charts do not have the data a good map does; it's no substitute. I applaud those pilots who still know, map in hand or handy, what is below and how far below it is.

Radar vectors for departing, en route, or landing, and the altitudes the person on the ground gives you are supposed to keep you from hitting anything. But don't count on them 100 percent. The people on the ground aren't 100 percent perfect, and neither are you and me. There have been accidents because an airplane was vectored into high terrain. Very few, but they have occurred.

In the final summing up, the person flying the airplane is responsible for it, and missing the ground is a paramount part of the job. People on the ground seem to take over more and more of our flying, but it's still basically the pilot's responsibility. Never take anything for granted, and especially not the idea that a certain heading and altitude given from the ground will clear all obstructions! The standard instrument departures in mountainous areas may not give safe altitude clearance if the aircraft's climb performance is lower than normal or if ice during climb has made climb marginal. Wind flow down mountain slopes can also reduce climb. It's well to look over the departure carefully in regard to terrain and not just assume that it's okay; and especially relate the departure route to your airplane's climb ability.

We must, by any means possible—omni, RNAV, ADF, DME, INS, Loran C, GPS, whatever—keep up-to-date on where we are. We must not simply sit there, wandering hither and yon following the ground-given vectors and not knowing where we are in relation to terrain. The controller may be giving headings, but the pilot is responsible for the navigation. Navigation is made up of many things, and one of the biggest is keeping position information and knowing where we are! The heading, whether provided by compass or ground-based ATC controller, is simply directional information. Where the aircraft is, and if it is clearing all terrain, I say again, is *always* the responsibility of the pilot in command!

17

Weather Flying
En Route

Now we level off at cruising altitude; sit back and relax a little. The intense concentration of an instrument departure is over. This is time to look around the airplane and check things. Glance at the fuel setup, pitot heat, engine heat, and deicing equipment. It's a good idea to start on one side of the flight deck and cover everything across the instrument panel to the other side.

The charts and papers from takeoff are probably mixed up, and this is a good time to get them organized and neat. Now there is an opportunity for writing down the takeoff time, the time a fuel tank was switched on, and for the other bookkeeping jobs. Knowing when fuel tanks were turned on and the time takeoff was made, along with the time over checkpoints, is important.

During all this, of course, we've been watching the weather. En route it's a clear-cut problem of being or not being in an area of instrument flying, with possible ice or thunderstorms. Turbulence can be mixed with either. It's natural, of course, in thunderstorm areas, but if it's in an icing area, the ice will be fast forming and heavy, and our problem is to get an ATC clearance out of there. Such a situation means unstable clouds, generally, and often a good way out is up. The clouds, probably embedded cumulus in a general overcast, will occasionally be visible as we bounce in and out of them between layers.

✝ Those Water Droplets

We hear a lot about icing and water droplet size, sometimes referred to as *micrometer*. The bigger, the more ice, which seems pretty obvious.

249

The point to make is that the NWS or FSS doesn't forecast or announce droplet size; the talk is all scientific stuff pronounced by people trying to impress other people. What this boils down to is you, the pilot, determining how much ice is up there; if things are turbulent, there will be moderate to heavy ice (big droplets). The same is true for weather that's intensified by upslope conditions and, of course, flight through fronts will have heavier ice, as will thick stratocu clouds in the lee of lakes. While the micrometer stuff is good cocktail conversation, it isn't any help to you. Simply being a good student of the current weather conditions—and whether or not it's turbulent, orographic, or frontal— is the way you decide how tough the ice will be and, of course, the forecast's prognostication. The task is to take your own look and superimpose your judgement on the forecast to assess what will be up there waiting for you and how it may change with your flight's route and progress.

✝ Think Ahead

We've talked before of the different weather conditions and how to cope with them, but here I want to talk about what we think about while flying from one place to another. Of course, we are navigating, listening to the radio, working ATC, and keeping track of ground speed and fuel usage to know if we'll get where we want to go with fuel enough remaining to go to an alternate if necessary.

As we do this, our background thinking is doing the important job of keeping ahead of the airplane. This is a tremendously important part of flying, in or out of weather, but especially in weather.

What do we do to keep ahead? We have the charts for the next leg of the trip in hand. As we fly toward one omni station, we have the course out from it well in mind before ever getting to the station. We know what the next station is, or intersection. We are mentally prepared for a number of courses ahead. We have the terrain and altitude clearance well in mind, too. No matter if we're using omni, Loran, or GPS, we should be heading conscious, anticipating, keeping ahead of the airplane. The time a fuel tank is due to run out is noted, so that we'll begin to think about it in advance and change it before it runs out.

But most important is the weather. We listen to broadcasts and copy sequences for our route, destination and alternate. It takes more than one hour's weather watching to catch the trend, to see if it's doing what the forecasts said it would do. If it isn't, then what does it seem to be doing? Is it getting worse or better? So we copy weather each hour and compare it with the last hour. It's wise, as we fly, to listen in on Flight Watch. Hearing weather broadcast to others, pilot reports, and requests can give an idea of what's going on in many areas. One can also pick up, by a feeling of urgency and certain tenseness in the back-and-forth conversations, what weather may be going sour and not acting as forecast.

Better weather, of course, isn't any problem. Worse means we have to plan for alternate action. What might that be? First, we should check deteriorating weather against the weather at our alternate. Is it holding? Can we feel confident that it is still the safe place to go if we cannot get in at our destination? If it is, we keep going to our destination, make an approach, and if we can't get in, we proceed to the alternate—just like the book says.

If the alternate is going down, too, then we need a new alternate. If our study of the big picture was complete before takeoff, any change from the forecast should be understandable. We should be well equipped to take alternate action. A good weather pilot has this alternate action in mind before takeoff, alternate action that fits the weather setup that day. Again and again, knowing the big picture is important!

Being ahead of the airplane means having plans made early to handle unexpected situations. Not being ahead is the person who faces a changed situation unprepared and in panic mode as to what to do next.

✝ What's it Like?

What's it like en route, as far as our flying condition is concerned? Well, we'll either be on instruments or not. We can be under clouds, on top of clouds, or between layers. We can be in and out of clouds.

Unsupercharged piston airplanes will be in cloud more than any other type, but even so, most "instrument" flight will be between layers or on top. Clouds rarely stack up from the ground to some great altitude. When they do, it is in a front or near the low center. These are not far across, nor do they stay in one condition for long periods.

A big portion of weather flying occurs in postfrontal conditions, and that means flying on top. Prefrontal conditions will be largely under an overcast that may gradually squeeze down on one.

What we are saying is that most instrument flight isn't on instruments. When it is, it's because we are flying in a front or low, as we said, or we've made a bad choice of altitude—like being in a stratocu deck when we could be on top of it.

This allows us to sum up and list the conditions. We've said we'll be either on top, under, or between. In these conditions we are staying out of ice, and we should be able to see thunderstorms and avoid them. When the clouds get together, we can get ice or fly into a thunderstorm. We are near the frontal surface. If we aren't, then we've just slipped into clouds that are higher or lower than the ones we've been flying through. If we understand the weather situation, then we should know whether to go up or down.

If the clouds have enveloped us and we're near the front, then perhaps it's time for ice. This isn't a reason to feel all is lost. Many times there are above-freezing temperatures at a lower level. Then it's simply a matter of getting a clearance to go lower, remembering terrain clearance.

If it's cold clear to the ground, then the job is to be certain that we are above the lower stratus clouds. This means a higher altitude. But in getting this higher altitude, we are not trying to get on top; we are just fishing for an area where there's little or no cloud, just snow.

We can pick the altitude above the lower stratus clouds and go into the front knowing we'll get ice and watching to see that we haven't gotten in so much that we cannot turn around and get out.

If we are crossing the frontal surface at right angles, it shouldn't be too far through; so let's be certain we are crossing it the fastest way. Let's not fly along the front; then we'd really ice up.

Thunderstorms we've talked about. Either they are air mass and we wander around them, or they are frontal and we are on instruments near them and need radar. Without radar or spherics, don't be in there! If the radar breaks down once we're in it, be prepared for a rough ride. Which points up that we shouldn't fly fronts—winter or summer—until we have a lot of instrument experience.

So en route we watch weather at our destination and alternate, but we also watch to see if any fronts are getting on the route. Generally, precipitation and easterly winds make us suspicious of a warm front. The frontal surface is out there somewhere where the winds go to southwest from easterly. Ahead of that is the large area of precipitation. That's where the weather is and where we'll find embedded thunderstorms in summer, and sometimes in winter, too. We'll find snow, and in a more or less narrow band, we'll find freezing rain. Until we're experienced, we simply run away from this area, either by making an end run if we have a long flight and this is possible, or by landing and waiting it out.

We pick up a cold front more easily, since its wind shift is more dramatic, SW to NW, with heavy rain, thunderstorms, or heavy snow and snow showers. It's rough in summer or winter, with ice in winter, thunderstorms in summer. Are you equipped to take it on? Equipped in skill and knowledge and airplane and gadgets? If not, beat a retreat and wait for it to go by.

Flying en route is the crafty time. Watch the weather, destination and alternate, and developments along the way that may mean something different on the move—big picture again. Watch the ground speed and its relationship to fuel and distance. These, together and interrelated, are the budget, and we cannot afford to overspend.

☨ Forced Landing

Suppose we are faced with a forced landing on instruments. What then? First consider the terrain under you. If there are mountains, turn so you descend parallel to them and not across. That way you have a chance of coming down in a valley and at least not banging head-on into the side of a hill.

On top of clouds one often can see a wavelike pattern in the clouds below. If so, descend in the low part, the trough of the wave, and you will probably be over a valley.

Slow down to minimum airspeed, but not so slow as to stall until ready to touch. Ninety-nine chances out of 100, you'll see something before you hit, and you'll have a chance to maneuver and pick a softer place. So you want a little maneuvering speed.

Full flaps and gear down. The gear will help to decelerate as it tears off, and the flaps will give the lowest speed, of course. With shoulder harness and pillows in front of you and your passengers, things may not turn out badly at all. The trick is to hit under control. It all sounds pretty desperate, but it isn't quite that bad. We hope it never happens to you, and it probably never will.

And, of course, en route flying is handling navigation and ATC. It is also, during the not-so-busy cruise portion, a chance to plan the arrival, to get out the arrival airport area charts and study them for routes and holding points. It's the time to study the instrument-approach plate and firmly fix in mind the lowest altitude we'll go to and what the missed-approach procedure is if we don't get it.

The approach plates should be held by a handy clip on the control wheel or instrument panel so that we can see them easily and without juggling them on our lap. With all this done, we are ready for descent and landing.

18

Landing in Bad Weather

As we approach the destination, weather takes on a more realistic feeling. Ceiling, visibility, wind, turbulence, and runway conditions become real because we are about to cope with them.

First we try to visualize the weather in relation to our descent. Is there ice or are there thunderstorms? If ice, as we've said, the job is to get well prepared (heat, props clean, etc.) and descend quickly to a landing before too much ice covers the airplane. Thunderstorms we have to detour around.

Now let's look back at the basics of an approach. There are two kinds: One is a precision approach with glide-slope guidance to the center of a specific runway, which is an ILS or MLS. The others don't have glideslope guidance or precise runway lineup; they are omni or ADF and called nonprecision. These approaches have higher minimums because the guidance isn't as precise. But because the minimums aren't as low, there's no reason for not taking them as seriously as an ILS. Now there are approaches made using global positioning systems (GPS). To date, these approaches are to higher minimums than ILS, but as the technology advances, GPS approach minimums will come down. Since ADF and omnis aren't as precise, we should do the best possible job so as to reduce their imprecisions to a minimum.

✝ Flying the Approach

Courses followed exactly and altitude controlled with precision are the rule. The altitudes on the approach plate are to be used accurately, and times from final fix to the field followed closely. Times for letdown should be modified according to the groundspeed we think we are making.

There are three phases of an instrument approach:

1. The all-instrument part.
2. The transition period from instruments to visual.
3. The visual part.

The instrument part is the easiest. This is a mechanical thing we learned when we got an instrument rating. But there are a few tricks and points worth talking over.

First of all, it's important to get the airplane settled down and ready for the approach, that is to get all the prelanding checklist items we can out of the way well in advance. We shouldn't be worrying about changing fuel tanks, getting gears down, etc., while we are trying to get on an ILS and stay there, track an omni, or home on an ADF. Have everything possible set for landing before the approach is started.

✈ The Instrument Part

The most important part is to keep on top of things every moment. We are talking again of scanning and keeping headings where we want them along with altitudes, descent rates, and airspeeds.

Let's remember the simple method of keeping a heading by watching our bank on the artificial horizon. If we never let bank go unnoticed, we will not get far off the heading. A part of this is promptly changing the heading when it should be changed. In other words, if you think a heading change is needed, make it now! The idea is to keep a short rein on the entire process and never let the airplane get far from the wanted path. Scan often and act quickly but, of course, smoothly.

Glideslope on ILS, or altitude change from an omni, is mostly a matter of descent rate. A localizer needle is not flown; the directional gyro is flown to a heading, and we refer to the localizer needle to see if that heading keeps us on course. In descent, the vertical speed indicator, hopefully an instantaneous type, is the key. You set up a certain rate of descent and see how that keeps you on the glideslope. If we go below, the descent was too fast, and so we get back on and try a slower rate. We bracket the glideslope using the vertical-speed instrument, just as we bracket the localizer with the directional gyro.

So the two important instruments are the gyro and the vertical speed: one for course control, and the other for descent. The horizon, showing attitude in roll or pitch, is the instrument by which we make our gyro turn or our vertical speed change. If a bank occurs, the heading will change and we stop it or bank to make a heading change. If the horizon shows our nose down more than it has been, we bring the nose back where we want it; if we want to go down, we make a forward movement which shows on the pitch part of the horizon; we should know about what pitch attitude it takes to come down a 3-degree ILS; it will vary, of course, depending on wind.

We refer to other instruments, too: airspeed to keep it within bounds and the altimeter to see we're where we ought to be. We often hear arguments concerning whether we control airspeed with the elevators and rate of climb with the throttles, or vice versa. Well, it's a silly argument because what we're managing is energy, and we do it the best way for the condition we're in. If we're below the glide path, our natural reaction is to pull back and get up to it. We zoom a little and get back, but if we're quite low and our zoom takes a long time, we will lose energy and slow down. Any experienced pilot knows this, and when pulling back automatically puts on some power. If it's a little pull up, and there is some extra speed, not much power needs to be added, maybe none. If slow and quite low, a lot of power may be added right away because pilot experience, subconscious or whatever, tells the pilot that additional energy is required.

But stopping to think which to do, add power or pull back, just isn't the way to fly an airplane. It's a smooth coordination of whatever it takes to get the job done under present conditions, and if a pilot doesn't understand that, a lot of practicing and learning is needed before trying to fly down an ILS looking for the bottom of the clouds at 200 feet or less! To do it all, we scan and scan often.

On an ADF approach, or omni, don't think something special is needed when the station is crossed. Often, after station passage, inexperienced pilots start making big turns and deviations. They aren't necessary. Approaching the station, we should have pretty well nailed down what the drift is and the required heading. After station passage, just apply that much drift to the next heading and cool it; wait a moment or two for things to settle down, and then resume tracking.

In a descent from an omni or ADF, we do not have glideslope. We simply leave one altitude and go to another. This doesn't mean we need to dive for it, but we should get on with the descent. Don't dawdle. The vertical speed shows us how fast we're doing this. Obviously, we don't want a descent rate that will make recovery at the desired altitude difficult. That's dangerous. What our descent rate should be depends on how much time it will take us from the final fix to the field. That says how long we have available to get down and that tells us the descent rate. If it's a minute, omni to field, and we have 500 feet to lose, a 500-foot-per-minute rate of descent will get us there on the button. But since we don't really know what speed we'll make station to field, we should add a little. If I had a 500-foot-per-minute descent rate required, I'd probably use about 700 fpm to be sure I'm down there before I get to the field.

The descent rate is also a guide for leveling off at the desired altitude. If the rate of descent is high, then we begin to level off farther above the altitude than we would with a low descent rate. We don't want to go below our target altitude. Be positive when leveling off. Don't delay. Pull the nose up and get it done! A little practice in hitting altitudes from various descent rates will soon let us know what our individual abilities are. It's good stuff to know before we start making approaches.

We want to get the approach tied down as far out as possible to reduce the need for excessive maneuvering close in. An approach is a little like a funnel. Well away from the runway, the localizer is wide, and as we get close in it becomes more narrow. This creates a tendency for us to be a little less precise out near the outer marker than we are in close. The way to make a good approach is to get on course and stay there as soon as possible. A good portion of missed approaches don't happen in the last part of the approach, although it looks that way; the miss probably started way back at the beginning of the approach when the pilot didn't get on track right away.

The wind will change as we descend, and this will affect our drift and descent rate. We will have to change headings and descent rates to keep up with it, but the earlier we have these things under control, the easier it will be to pick up a change. And remember, the descent rate to stay on the glideslope is a clue to shear.

What we are trying to do all down the approach is to keep the changes as small as possible. Close in, the ILS is narrow, the omni station small, and small corrections do a lot. It's the time and place for maximum concentration.

✝ Close in, Things Get Tight

There's a lot of pressure when we get in close on the approach; everything becomes tighter and tighter and we must watch each instrument intensely and scan the important ones frequently. This is the place where hands get sweaty. It's the place, too, where the pilot has to be relaxed but alert. We must feel like a good athlete, intense in concentration but smooth in coordination and relaxed enough so that we move freely. This comes by practicing, by getting the approach under control early, and by talking oneself into the right mental attitude.

✛ Stick with It

As we anticipate breaking out and seeing the ground, we should remember one of the most important parts of an instrument approach. It is this: stay with it!

All approaches are not to the minimum altitude before seeing the ground. Most of them are not. The ground comes into view in various ways: through breaks in clouds, or a clean break out of cloud but into poor visibility, perhaps straight down in snow. We often see the ground before we have sufficient visibility to see the airport or runway.

But seeing the ground, good old Mother Earth, whom we may not have seen for hours, doesn't mean the approach is finished. It isn't finished until we land and turn off the runway.

Inexperienced pilots may attempt to continue the approach visually, thinking they know where they are, but with poor visibility and the field not yet in sight, things look different. Even an area one thinks one knows well will suddenly be as foreign as the moon when the visibility is only straight down! About then the ground may disappear and that's a profound shock!

During ground contact navigation, the pilot probably has wandered from the radio aid direction, but now, with visual reference suddenly gone, there's a wild attempt to get back on course with radio aid, but it's too late and the approach is missed.

When a pilot leaves the radio aid and begins to navigate visually, a sneaky descent probably begins. It's a natural reaction to get lower and have a better look, or it's done because one isn't scanning back to the instruments enough to check that altitude and airspeed are in safe territory. This pilot has entered a very dangerous domain!

The altitudes for an instrument approach are set up to clear all obstructions along the approach path. It is best, by far, to follow these altitudes precisely and not be "suckered" into letting down below them because one can see the ground and get the false impression that this makes everything okay.

If there is enough ceiling, the approach altitude will put you in the best position to land. If there isn't enough you shouldn't be landing anyway. Why deviate from the procedure? There isn't any good reason, so don't. I cannot stress this enough. Whatever the approach, it is designed to lead us to the runway at the proper altitude, so why deviate from it? No, hang right with it until landing is assured.

All this is especially important at night, even in fairly good visibility. The lack of a good solid reference can cause sensory illusions that will lead a pilot into the ground when thinking all is well. A few lights or, in daytime, a dimly seen building or tree clump, combined with a wing being down and off-level nose position, can easily give the illusion that one is higher than it appears. Lots of airplanes have piled up because of this.

When the runway or approach zone slopes up, you think you are higher than you really are. When runway lighting is set to an intensity lower than normal, you think you are higher than you are. Snow cover does the same thing, and under conditions of haze, smoke, and darkness, you think you are higher than you are.

These sorts of illusions have caused accidents. The answer is to trust instruments, not what you think you see. Stay with the instrument approach procedure until the runway is in view and it's definite that you "have it made."

This is the dangerous place, when one casts aside instrument reference and uses only one's eyes for guidance. Again, and again, we must not descend below the final altitude until the runway is well in view; we can adjust our descent by reference to it, and we're certain of our landing. Even at the point when we cross the runway's end, we should be checking instruments—airspeed and especially altitude. Even with the runway in view, it is possible to get too low, because our eyes play tricks; and a look at the altimeter will sometimes be startling! But if one has taken frequent, quick looks inside at the altimeter during descent, even though "contact," these surprises will not happen.

During an ILS approach, keep in mind how much crab angle has been necessary to stay on course. Let's say the ILS course is 50 degrees and it has taken about 60 degrees to keep on the localizer. This means that when we break out and see the runway, we'll have a 10-degree crab angle. The runway will come into view out the left side of the windshield. The instinctive reaction is to kick this crab out and line up with the runway. We shouldn't because the airplane will then drift off the centerline, and it's surprising how fast this happens. If the ceiling is low, we'll probably not have time enough to maneuver around and get back on the runway before we reach it. Don't attempt to get lined up with the runway once you are off to its side by any appreciable amount. It's a case of, "You can't get there from here." The geometry and dynamics prohibit wild maneuvers that will get you back on the runway, unless there is only a small correction needed. It is time to swallow pride, pour on the coal, and do a missed approach. It's a lot better than hooking a wing or pod and maybe crossing the runway awkwardly to wind up in the boondocks.

So on the ILS, note the drift and visualize where and at what angle the runway will come into view. Then don't be too quick to kick out the drift. See how you are doing first. You may have to hold it all the way to the ground.

✝ When We See Again

Now we've made the approach and we break out where we should. What happens? Unfortunately, there's a great tendency to push over and "go" for the runway. They call this the "duck under" maneuver. What it does is put one too low, and it's dangerous.

There's more than one way to "duck under." So-called experts blame pilots for shoving it over to see the ground. Well, there are probably some cases of that nature, but I have more faith in pilots' intelligence and will to live. Another reason for the duck under, as we've said, is flying contact and forgetting to scan the instruments frequently because, basically, even though we see the ground it is still IFR flying in low visibility.

But there is another reason for getting below the glideslope, and that's shear. Normally as we descend to land, the headwind becomes less during the descent. This causes airspeed loss and therefore the aircraft's nose to lower, which, if we're not right on top of it, will cause us to slide under the glideslope either on instruments or trying to sneak in contact. It's one of the main reasons we're startled to see an altitude loss when we look back at the altimeter. This isn't dramatic shear, but simply a few knots of wind reduction during descent. It's a subtle thing, but one to watch for. It emphasizes, again, the importance of scanning those instruments.

To go back a bit, we were on instruments following the glideslope and we established a good rate of descent. But the moment we look up and see the approach lights, we lose glideslope information. The approach lights don't give any. The radio glideslope is inside on the instrument panel, not outside where we are looking. Unless there's a visual approach slope indicator (VASI) we haven't any glideslope reference. Looking back inside at the glideslope indicator doesn't help because the glideslope deteriorates as we get lower.

ILS systems approved to CAT 11 minimums, generally 100 feet, have accurate glideslope indication almost to the ground, but others are not that precise. Their glideslope accuracy deteriorates as they approach the ground. How high this happens depends on various things, but terrain has a lot to do with it. Some glideslopes are good to 50 feet, but I've seen others that dip or become erratic as high as 150 feet. So at low altitudes if the glideslope isn't a CAT 11 and the needle suddenly makes a fast and pronounced excursion off course, don't chase it. Simply hang on to the rate of descent that's been taking you down the glideslope successfully and use that for a glideslope.

This is a dangerous area. Enough airplanes have landed short under these conditions to prove the point. The airplane condition at that point aggravates the situation. It is going slowly. The pilot pushes over. The airplane goes through a shear zone. It sinks and sinks to a dangerous level before the pilot can catch the cue visually, if ever. What we need is VASI or PAPI (precision approach path indicator) on all instrument runways, but we don't always have them. There are variations of VASI, and one should study and know them all. Equally important is to study the various lighting systems for approach, runway end, and runway lights before doing serious instrument flying and approaches.

Gaining in use is heads-up display, where the necessary guidance is on the windshield—really good stuff. I flew many approaches with it in Air Inter's simulator—some to zero, zero with go around; it is a beautiful, simple way to fly. Air Inter is a French airline with a marvelous weather record using a heads-up system.

But what do we do about it now? First, remember the rate of descent you've been carrying in the approach. When you break out, take occasional quick looks at the instrument to see that you are holding near this rate. If you are, the descent path should be nearly correct unless there is bad shear. Check altitude, too, especially in relation to the middle marker, the amber one. On the approach plate, there is a published altitude for crossing this marker. It will be around 200 feet AGL. Be certain not to be below this middle-marker altitude until after passing it.

What we've said in all this is not to trust your eyes for glideslope guidance when visibility is low, especially at night. This applies to the VFR pilot under fairly good conditions, too. We've also said that an approach doesn't end just because we can see the ground. And when we see the ground, we must kill the desire to push the nose down and duck under!

✝ Autopilots Doing the Work

Many aircraft have autopilots that make the approaches and do it magnificently. But this doesn't relieve the pilot from constantly monitoring how the autopilot is doing. It's important to know the heading it's flying, descent rate, and power if there are auto throttles. It will be very handy if the autopilot should suddenly disconnect or fail, and this does happen. And, of course, it's a check that the autopilot is doing its thing properly. Careful observation of the approach then has the pilot informed and ready to take over manually, for any reason, with minimum fuss.

✝ Circling to Land

Sometimes an approach is made but the wind is on a different runway and it is necessary to break off and circle to the other runway. This is a dangerous procedure and requires higher minimums, generally 1,000 feet or more. In rain or snow, with poor visibility, perhaps at night, this is not easy. The tendency is to look out at the ground and runways, forgetting the instruments inside. It is very easy while doing this to have altitude slip away unnoticed and, with bank giving illusions, get far too low. How to do it? By instruments! Fly headings and maintain altitude with occasional looks outside to see how you're doing until lined up on the heading of the desired runway and you can see it for landing. It is a watching in and glancing out procedure with constant instruments reference necessary. If at all possible, don't do a circling approach under poor weather conditions!

✝ To Touch the Ground

So we've made a proper approach and now see the runway. What's next? Actually, we continue with the descent and approach problem. We cannot get too low until the runway pavement is under us. While we want to get on the ground, we also don't want to try to cut it so short that we leave the wheels on the last row of approach lights. It's one of aviation's many compromises in that you don't want to land too long with the danger of going through the fence at the other end, and you don't want to land too short either.

This points up an important point about landing the airplane once the runway is securely under it. The point is to get it on the ground! This isn't the time to float along with a little excess speed trying to gently slide the airplane on the ground and impress those riding along. Even if the landing is a little rough, get it on the ground where the wheels and brakes can begin to get it stopped. And in doing this, we want as much runway ahead as possible in case we start aquaplaning on a water-covered surface or sliding on ice.

So there's a nice, delicate balance required in the approach. It starts with airspeed before the runway. We need enough to take care of gusts and shear, but not so much that we come screaming in over the runway and float halfway down it before we touch down. But we don't want to land so short as to risk hitting the approach lights, or we don't want to land so long that valuable runway is flown over instead of being used for getting stopped, These are the kinds of judgments pilots make alone and without help.

✢ Low Visibility

Now a few points about low visibility. It's interesting to note the following table of the percent of missed approaches versus visibility, as taken from records at London airport for three years:

Runway Visual Range
Missed Approaches

(Feet)	(Percent)
1,970-2,300	22.2
1,620-1,970	30.6
1,475	40.5

It simply says that it's tough to make a successful landing when you cannot see. It says there isn't any point to being a hero trying to get in under conditions that are just too tough.

These low visibilities are generally there because of fog. During this condition, the visibility-measuring device may only see a portion of the runway area, although some runways have RVR measurement in two and possibly three areas along the runway and give rollout area visibility values. Beyond that, where one may roll out, there could be a zero-zero fog patch. Very low visibilities are touchy and require objective analysis about whether or not it's worth trying before doing so.

✢ Ground Fog

There's another fog condition that deserves attention, and it doesn't have to do with an instrument approach. We can fly in clear weather, day or night, and note that our destination reports zero in ground fog. We get over the field and are surprised to see the runway and airport below us well in view.

This is simply because we are looking down through a very shallow layer of fog, rather than through it horizontally. But if we say, "Heck, I can land in that," and make an approach, we'll get an awful shock as we descend and suddenly, about the time we start to flare, go on solid instruments.

What then? Pour on the coal and get out. If it's too late for that, hold a steady descent rate and attitude until you touch. Then get stopped as fast as possible and hope you don't hit something before you stop.

✝ On the Ground

Back to our low approach to an airport. We've gotten on the ground and stopped; is the approach over? No. The next order of business is to quickly clear the runway. There's an airplane at the middle marker about to land, and the tower cannot clear it until you're off the runway. So turn off as soon as possible.

In doing this remember that the so-called high-speed turnoffs aren't as high speed as they sound. If the runway is wet and we start to turn off at fairly high speed, we can be shocked to find that the nosewheel may have turned toward the taxiway, but the airplane hasn't. The nosewheel, without much weight and traction, is just skipping sideways while the airplane is going somewhere between the runway and taxiway, which is in the boondocks! Find the speed your airplane can make that turn on a slick surface, and be down to it before trying to turn off.

✝ The Toughest Case

Suppose the worst happened and for some difficult-to-explain reason one got caught with no alternates, no fuel, and a fog-covered, zero-zero airport the only place to go. I did this in Alaska on the Aleutian chain during World War II. It wasn't zero-zero, but it might as well have been. The ceiling was essentially zero, and the visibility was about 200 feet in snow and fog. Fortunately, I was over a good airport, Shemya, which had a long and wide runway.

I came down the ILS and carefully checked the descent rate needed to follow the glideslope. I flew the ILS as tight as possible. When we were under 200 feet and the glideslope started to wiggle and deteriorate, I just held the rate of descent until we touched. Then, control wheel forward (taildragger B-17), I slammed on the brakes and tried to stay on the localizer and get stopped before hitting something.

We made it. Incidentally, I made a few practice approaches and go arounds to get an idea of drift and descent rates before doing the one for keeps.

What really impresses you is that with near zero visibility, you really don't see! A few fuzzy lights aren't enough for guidance, and you are helpless and feel helpless. This feeling can occur with visibilities up to 1,000 feet or so, depending on how fast you land. It's an excellent lesson to make one realize that there are human limits and we shouldn't try to exceed them. Too many things happen too fast.

I've made many approaches in a 747 to 100 feet with 400-meter visibility. It's done on automatic pilot, and it is the best way to go because you have time to check all phases of the approach constantly, while a bunch of transistors and servos do the manual labor. Without this capability, a pilot flying very low approaches by hand either will be concentrating so hard on flying that there will not be time to check other important things like altitude and instruments that tell the flight's entire condition, or, if those things are being checked, there will not be time to keep on course and complete the approach. It cannot be done unless the pilot is a superperson, and while I've known the best of pilots, none of them are superpersons. One reason they are the best is that they know when to admit something cannot be done safely!

The automatic approach right down to landing has proven itself because accidents don't seem to happen when doing it this way. The last time I checked, there hadn't been any on automatic zero-zero landings! This seems to prove another interesting point I've been preaching. If folks stay with the approach right to the runway and overcome that desire to finish it by eyeball rather than by instruments, there will be a lot fewer accidents.

19

Teaching Yourself to Fly Weather

With the idea that one crawls before walking, we can teach ourselves to fly weather. It's a progressive, self-taught process that uses our own program.

In the work of getting an instrument rating, we made practice IFR cross-country flights, getting clearances, making radio reports, navigating, gathering weather, and all the rest, even though we weren't flying actual bad weather. Now it's time to start our "home study" weather program.

Each day, in our advancing times, the complexities of air-traffic control, routes, and communications grow, so that all the experience we can get in this area is important. If, on each flight, VFR or IFR, we are on a flight plan, doing all the work required, we will become facile with this part of the job and do it smoothly, almost automatically. Once this has become an easy task, we will have time to think about the weather.

✠ Where's the Emphasis?

The balance in flying between ATC and its communications versus the actual weather has tipped toward ATC as the major problem. I am certain that weather accidents have occurred because pilots became so engrossed with ATC and some altitude or route that ATC gave them that they got in trouble with ice or thunderstorms because their attention wasn't on the ice or thunderstorm as the primary problem. We tend to be more afraid of authority than of weather.

This, of course, is illogical and perilous. If the route or altitude assigned doesn't fit, it's time to tell ATC that you want something different. If we are battling a difficult situation and ATC keeps pressing us with complex routings or unreasonable requests, we should be prepared to tell them that we have urgent weather problems and need help, not hindrance. They will help if at all possible.

All this comes easier if we are prepared by being able to handle the ATC complexities, by having a well-organized cockpit, by knowing the regulations, and by having a boom mike to make transmission a simple matter that doesn't require taking a hand from the flying job. It is difficult to find just the weather we want for practice, but ATC is always there for practice. We should do as much as possible.

☩   Learning the Weather

Now let's get on with learning to fly weather. The actual weather part we will sneak up on by flying a little at first, more as we gain experience.

Following is a step-by-step method. These steps are guides and one's own judgment will vary them as one appraises his or her growing ability and degree of comfort in different stages of weather.

The idea is to fly weather with safeguards that relate to our experience. After we've flown the first step's conditions enough to feel comfortable, we can take on a little more as in step two, and so on. The steps are:

1. Fly good weather to good weather on top.
2. Bad to good.
3. Good to bad.
4. Bad en route.
5. Thunderstorms.

   Now, let's talk about them.

Step One

The first step, *good to good*, means we leave a point that has broken clouds or better and fly toward a point that has the same condition and is forecast to remain that way or to improve. The ceilings should be 2,000 feet or higher, the tops 7,000 feet or lower. No fronts should be moving toward the destination. Temperatures should be above freezing, and there should be no thunderstorms.

This flight will mean climbing up through clouds to on top and descending through them at destination. It will require weather watching to be certain things stay as we want them, and, of course, a flight plan and ATC.

This condition will generally be found in the stratocu cloud decks behind a low passage. It gives an opportunity to take on more or less weather, depending how close behind the low, or front, we take off.

At first we might wait a day after the low passage for the stratocu deck to become thinner. Later, as we gain experience, we can take off closer and closer to the departing weather until we are taking off with a ceiling that's quite low, climbing to a top that's quite high, or flying on instruments to a destination that's overcast.

Step Two

This leads into step two, *bad to good*. It's simply a continuation of the first step, but departing closer to the low and front until we are taking off just after its passage. We should be careful not to take off before the front has passed or while it's passing, because this can be a very demanding experience.

The key thing in both these steps is that we are always flying toward good weather. We want to be certain it's forecast to stay that way, too—no fronts moving in, as there might be if one took off from an airport that just had a warm front passage with weather improvement and then flew across the warm sector of the low to come upon a cold front. In this case one should realize that out west somewhere there's bound to be a cold front coming along that we don't want to get involved with. When starting these first steps, it's best to take off after a cold front has passed. Then there shouldn't be any more fronts for quite a long distance.

There are special situations, like the Los Angeles basin, that are excellent for bad-to-good flight experience. The frequent low stratus allow for an instrument departure and a climb to on top, where it's CAVU, and then a flight to someplace on the desert like Palmdale where the weather is good. And generally this can be done without running into the problem of fronts.

We can back up and start these two steps over, but with some ice. The first sight of a smear of ice across the windshield is quite interesting if there's a way out nearby. We can do this by climbing through a thinish, below-freezing stratocu deck, 4,000 or 5,000 feet thick, that has a good base, 2,000 feet or better, at both departure and destination. As we get more accustomed to ice, we can take this deck a little thicker and a little lower. (The airplane should be certificated for flight in known icing conditions.)

A good method to learn about ice is to fly in a stratocu deck that has ice and at the same time a freezing level at some safe altitude below, so one has the chance to go down and melt any ice. The freezing level must be above the ground enough for 2,000 feet clearance or so. This condition is often possible in the spring and fall.

Step Three

Step three is *good to bad*. This means to fly toward a destination that has weather approaching. The departure should have solidly good weather. Doing this, we can go to an area of low ceilings, shoot a low approach, and if there's trouble, turn around and go back to where it's good.

Plan, at first, to get to the destination well ahead of the weather, but be prepared for surprises. It may move faster than expected. You'll learn, this way, how foolish it is to fly to a destination that calls for deteriorating weather, without a good out to run for.

As experience is gained in this step, one can fly to destinations with worse and worse weather. The key is to have that out; in this case good weather behind. This step should be tackled first without any ice, and then we gradually work into ice as we've done in the previous steps.

Step Four

Step four is *bad en route*. This means we'll take on a situation that has a good destination but something in between that is tougher than just a stratocu deck. It will be some kind of front.

To start this we should be without ice and thunderstorms, just as we begin each step. The destination should be forecast good and to stay that way or improve. Any front should be well past the destination before we take off. The takeoff point should be 1,000 feet or better and forecast to stay that way for at least two hours after takeoff.

We progress in this step by taking off with the weather closer to our departure point. This requires thinking about a takeoff alternate—where to go if you must return and the takeoff airport has gone below limits.

We develop this step until we are taking off with bad weather, flying through it en route, and landing at a point that has recently cleared. At this point you have reached a pretty sophisticated level in weather piloting. But there still remain thunderstorms.

Step Five

To begin *thunderstorms* we should fly with only air-mass types en route. They should be forecast as scattered so that we can wander around them and look them over from a safe vantage point. There shouldn't be any fronts forecast within 500 miles of our route or destination. We should have lots of fuel. The destination may be covered by a heavy shower when we get there, and we want enough fuel to wait it out or go on to a thunderstorm-free area. Plan to arrive at the destination with lots of daylight remaining, three or four hours. Don't go on instruments with thunderstorms in the vicinity. Don't try to top them. Don't cut too close between two of them. Don't fly under the anvil overhang.

The progression of weather learning in thunderstorms is difficult because any thunderstorms beyond air-mass types will be frontal, and that may mean getting on instruments. Don't do this without radar and knowledge of how to use it.

We could fly under high-level warm front thunderstorms as a progressive step, but this is flirting with heavy rain and low stratus clouds that mean instrument flight. After air-mass thunderstorms, therefore, the next step is a big one, because one may be forced to fly through a thunderstorm.

Flying thunderstorms in fronts and lows is the big time, and unfortunately has to be learned the same way swimming sometimes is, by diving in over your head the first time. A person should at least know how to fly turbulence before doing it. Radar, as always, is a must. And this, again, isn't to help you fly through thunderstorms but is just an extension of your tools for keeping out of them. A spherics system will also be a big aid to avoid thunderstorms in what we might call a "wide" sense. What we're saying is don't go into a heavy thunderstorm area, on instruments, counting on spherics to weave you through as you would with radar.

The learning steps I've outlined are not hard-and-fast rules. They are guides to start from and think about. But there are a few firm points that should be rules.

1. Always have an out. Fly toward good weather or from good weather that you can return to.
2. Take on ice only after other experience has been gained.
3. Don't fool with thunderstorms for a long time.

4. Don't get on instruments near thunderstorms without radar or a spherics preview.

5. Have lots of fuel.

6. Have lots of daylight remaining after the ETA.

There are other rules, and they are in the book in many places, but if I summed up one weather rule it would always be: *Have an out! And know the big picture!*

Looking back over these steps, we can see that weather flying experience isn't gained quickly. We need several seasons, years, to see the things we should see and experience the things we should experience. We must face the facts of weather flying. It cannot be gotten by injection, it cannot be gotten by reading a book, it cannot be gotten quickly. We must remain humble for a long time and know when to quit or when not to go. An instrument rating is a beginning, not an endorsement that one can fly off in any weather.

During the learning of weather flying, we should be careful about reverting to VFR flying. To duck down low and try weaving in and out of mountains or even over flat terrain with minimum visibility is a sure way to trouble. Remember, VFR in marginal weather is the most dangerous way to fly!

20

Something on Judgment

It is important to realize that in weather judgment, more determination is required to sit and wait than to fly. The press of wanting to get somewhere will overshadow the gumption it takes to drag the luggage back to town, miss a trip we really want to make, or delay getting home. But these reasons, and more like them, have caused many accidents.

In judging weather before takeoff and then skillfully assessing and flying it once airborne, our attention and action should not be influenced by other factors unless they become a part of the judgment, such as a mechanical problem that requires an immediate landing.

As I said before, it is not necessary to remain riveted to the idea that we must land at the airport of desired destination. There shouldn't be a fixation on that. Weather changes and conditions may well make it prudent to go elsewhere, and personal and emotional influences should not prevent tossing the desired destination aside in favor of a good, safe one.

✝ Limitations

The technical world of our day has put a lot of impressive equipment in airplanes to combat weather and make flying tempestuous stuff easier: good radio and navigation gear, deicers and anti-icers, radar, spherics, auto flight systems that take us to zero-zero, instrument systems to help us fly shear, performance, range, antiskid for braking, reversers, systems to keep windshields clear, flight directors to think for us on ILS approaches, instruments like HSIs that pictorialize, flight management computers, and glass cockpits that think, at least to an automaton's capability. There is more coming.

But first, let us clearly and forcefully remember that all this equipment will not fly all the weather. Mother Nature will periodically dish out weather that we simply cannot manage.

We've gained much—the ability to land with no visibility if the airport and airplane are equipped. We can top much of the weather we once bounced in and worried about. There's a long list, but no matter how superbly equipped an aircraft is, the inside of a thunderstorm is still harrowing, and severe shear on landing, instruments or no, can do the airplane in. Running low on fuel for many reasons can menace flight. Landing in a hurricane can, too. Periodically we are presented with weather of a most savage kind, and saying we've reached the age of "all-weather flying" is akin to the folks who said the Titanic was "unsinkable."

With our equipment has come the need for pilots to use it properly. Program computers with the correct numbers. Keep close watch that all is working as desired and the aircraft is headed in the right direction, high enough to miss all terrain. This doesn't come easily because the sleekness of the airplane and its near perfection lull one into feeling that nothing can be wrong. Here's where that overused but potent word, complacency, reaches its peak. The fact is that our modern aircraft demand more attention, not less, plus plenty of that old-time pilot's belief that things can and will go wrong and you'd better have a skeptical eye roving the cockpit and weather on a regular basis. The equipment and smoothness of aircraft, the quiet shirtsleeve environment, and digital readouts flicking away easily give a false sense of security, which tends to overlook the basics. How very treacherous this false sense of security can be!

All this has sounded like talk of modern airline and corporate aircraft only, leaving the little people out. But not so. The most humble aircraft now can have Loran C, HSI, GPS, spherics, and many other benefits, so the person flying it can also be affected by a false sense of what the airplane can do. It would be wise to remember that while well equipped, the airplane isn't an airliner or expensive corporate jet, and in most cases the pilot isn't as well trained as the airliner pilot, nor is he assisted by a copilot.

We have also been affected by the complexity of the ATC system and the bureaucratic stifling of aviation that has required and produced training of a rote nature that misses out on judgment. We all wish there was a way to teach judgment, but it is difficult. Some people seem to have it naturally—horse sense—and a feeling for weather that intuitively tells them, "it just ain't smart to go." Others develop judgment by experience as "fright time" is accumulated.

In flight, weather often takes a back seat to the system, the system being ATC with the pressure of fast talk, keep 'em moving, land one right close after another, and don't miss any instructions or make errors. With this is the frustration of delays, holdings, long vectors, and reroutes.

In the area where most accidents occur, approach and landing, there's an element of being too engrossed in the fact that one airplane is landing right after another. We're part of that and don't want to interrupt it, so we continue our approach sort of mesmerized by the idea of landing, without really thinking weather. The pilot ahead landed, why shouldn't I? It's a form of competition, also. We may not be willing to be the oddball, the one who couldn't make it and had to pull out, disrupting everything. Unfortunately, it takes courage to break out of the "norm," to say there's a thunderstorm too close and I'm going elsewhere, or the crosswinds are too strong and I want a different runway. But we must remember that weather is the first priority—not ATC, not the airplane ahead or the airplane behind, or the fact that ATC may send us off to Lord knows where if we want to stray from the routine flow. But the law of preservation and the rule of safety first must be the way. Despite all modern equipment, weather will, from time to time, humble the structure of aviation. The pilot's task is to be aware of these times, responsive to them, and have enough grit to take action. This applies whether the pilot has 30,000 hours or 10.

I was fortunate to fly with pilots who showed me what command meant. We were flying DC2s and DC3s then. On one of my first copilot flights, the captain was Jim Eischeid, a veteran with a special reputation for being an excellent weather pilot.

We were going to fly Flight 7 from Newark to Chicago. It was a rotten night with Chicago forecast low, ice en route. I stood next to Jim in the dispatch office, discussing the weather with the meteorologist, somewhat in awe, anxiously awaiting the flight to see how this man, whom no weather could stop, would handle it. Imagine my surprise when he turned to the dispatcher and said, "It's no good. I cancel!" He knew when not to fly. Jim retired after a long career that went from open cockpit mail planes to Connies, and he never scratched one of them.

Now we fly in an era that comes closer to all-weather flying. We can fly more weather safer than we did back then and, after all, we want the airplane to work and deliver people and goods reliably, but we have not truly reached the stage of total all-weather flight, from general aviation to airline, and we probably never will. There will always be some time when it will be wise for a pilot to say, "I cancel," or "I'm diverting!" The thing is to know when it's that time for each of us, and then have the guts to do it!

Suggested Reading

Anderson, Bette, and Roda, *Weather in the West*, Palo Alto, Calif., American West Publishing Co., 1975.

Bradbury, Tom, *Meteorology and Flight*, London, A & C Black, Ltd., 1991.

Byers, Horace Robert, *Synoptic and Aeronautical Meteorology*, New York, McGraw-Hill, 1937.

Chaston, Peter R., *Weather Maps: How to Read and Interpret All the Basic Weather Charts*, Kearney, Mo., Chaston Scientific, Inc.

Dunlop and Wilson, *Weather and Forecasting*, New York, Collier/Macmillan, 1987.

Edinger, James G., *Watching for the Wind*, New York, Doubleday, 1967.

Federal Aviation Administration, *Aviation Weather*, Washington, D.C., Government Printing Office.

Lester, Peter F., *Aviation Weather*, Englewood, Colo., Jeppesen Sanderson, Inc., 1995.

Petterssen, Severre, *Introduction to Meteorology*, 3d ed., New York, McGraw-Hill, 1968.

Wallington, C. E., *Meteorology for Glider Pilots*, London, John Murray, Ltd., 1977.

Index